PERGAMON INTERNATIONAL LIBRARY
of Science, Technology, Engineering and Social Studies
The 1000-volume original paperback library in aid of educatio[illegible] industrial training and the enjoyment of leisure
Publisher: Robert Maxwell, M[illegible] C.

THE NUCLEAR APPLE AND THE SOLAR ORANGE

Alternatives in World Energy

Other Titles of Interest

DE MONTBRIAL
Energy: The Countdown

DE WINTER
Sun: Mankind's Future Source of Energy

EGGERS-LURA
Solar Energy for Domestic Heating and Cooling

EGGERS-LURA
Solar Energy in Developing Countries

GABOR et al.
Beyond the Age of Waste

GRENON
Future Coal Supply for the World Energy Balance

GRENON
Methods and Models for Assessing Energy Resources

HOWELL
Your Solar Energy Home

HUNT
Fission, Fusion and the Energy Crisis, 2nd Edition

McVEIGH
Sun Power

MURRAY
Nuclear Energy, 2nd Edition

SECRETARIAT FOR FUTURES STUDIES
Solar Versus Nuclear: Choosing Energy Futures

SIMEONS
Coal: Its Role in Tomorrow's Technology

SIMEONS
Hydro-Power

SIMON
Energy Resources

STARR
Current Issues in Energy

UNECE
Coal: 1985 and Beyond

*Pergamon Related Journals**

Annals of Nuclear Energy
Energy
Energy Conversion and Management
International Journal of Hydrogen Energy
Progress in Nuclear Energy
Solar Energy
Sun at Work in Britain
Sun World

**Free Specimen Copy Gladly Sent on Request*

THE NUCLEAR APPLE AND THE SOLAR ORANGE

Alternatives in World Energy

MICHEL GRENON

International Institute for Applied Systems Analysis
Laxenburg, Austria

Translated from the French by

R. PENISTON-BIRD

PERGAMON PRESS

OXFORD NEW YORK TORONTO SYDNEY PARIS FRANKFURT

U.K.	Pergamon Press Ltd., Headington Hill Hall, Oxford OX3 0BW, England
U.S.A.	Pergamon Press Inc., Maxwell House, Fairview Park, Elmsford, New York 10523, U.S.A.
CANADA	Pergamon of Canada, Suite 104, 150 Consumers Road, Willowdale, Ontario M2J 1P9, Canada
AUSTRALIA	Pergamon Press (Aust.) Pty. Ltd., P.O. Box 544, Potts Points, N.S.W. 2011, Australia
FRANCE	Pergamon Press SARL, 24 rue des Ecoles, 75240 Paris, Cedex 05, France
FEDERAL REPUBLIC OF GERMANY	Pergamon Press GmbH, 6242 Kronberg-Taunus, Hammerweg 6, Federal Republic of Germany

British Library Cataloguing in Publication Data

Grenon, Michel
The nuclear apple and the solar orange. -
(Pergamon international library).
1. Atomic power - Economic aspects
2. Solar energy - Economic aspects
I. Title
333.7 HD9698.A2 80-40836

ISBN 0-08-026157-4
ISBN 0-08-026156-6 Pbk

Printed in the United States of America

CONTENTS

ABOUT THE AUTHOR

MICHEL GRENON has devoted practically his entire career to problems of energy: in particular at the Institut Français du Pétrole, the Atomic Energy Commissariat, and then at Euratom, where he was Deputy Director of the Petten centre (Netherlands), which he helped to set up. For ten years, making use of his experience in the fields of oil and nuclear power, he has acted as adviser to major international organizations and associations, and he participated in the Energy Policy Project of the Ford Foundation, at the same time giving courses at an American university on world energy policies. In 1974 he joined the International Institute for Applied Systems Analysis in Laxenburg, Austria, a scientific bridge between East and West, where he is concerned with "systems" aspects of energy resources and other natural resources, as well as with methods of estimating world oil resources, a highly controversial subject. He regularly contributes to numerous special interest magazines in France and abroad, and he has published a number of books on energy problems, which have been translated into various languages.

LIST OF ABBREVIATIONS

Organisations:

API	American Petroleum Institute (USA)
CIA	Central Intelligence Agency (USA)
EDF	Electricité de France
IAEA	International Atomic Energy Agency
IFP	Institut Français du Pétrole (France)
MIT	Massachusetts Institute of Technology (USA)
NASA	National Aeronautics and Space Administration (USA)
NEA	Nuclear Energy Agency (of the OECD)
OAPEC	Organization of Arab Petroleum Exporting Countries
OECD	Organization for Economic Co-Operation and Development
OPEC	Organization of Petroleum Exporting Countries
WAES	Workshop on Alternative Energy Strategies (USA)
WEC	World Energy Conference

Types of nuclear reactor:

AGR	Advanced Gas Cooled Reactor
BWR	Boiling Water Reactor
PWR	Pressurized Water Reactor

Various:

bbl	barrel (of oil: equivalent to 159 litres)
D	Deuterium
GNP	Gross National Product
NPT	Non-Proliferation Treaty
ppm	parts per million
T	Tritium

UNITS

The units most frequently used in the energy field are the TCE and TOE for energy balances and resources, and the MW and the kW.h and their multiples for units of electricity.

TCE	Tons of Coal Equivalent (7,000 kilocalories per kilogram or kcal/kg)
TOE	Tons of Oil Equivalent
kW.h	Kilowatt hour. Also used is the TW.h = 10^{12} watt hours = 1 billion kW.h
MW	Megawatt = 10^6 watts

The values ascend as follows:

Watt
Kilowatt = 10^3 or 1,000 watts
Megawatt = 10^6 or 1,000,000 watts
Gigawatt = 10^9 or 1,000,000,000 watts
Terawatt = 10^{12} or 1,000,000,000,000 watts
Therm = 1000 kilocalories = 1,000,000 calories

The TW, like the watt from which it derives, is a unit of power. The corresponding quantity of energy is the TW-year: frequently the term TW-year per year is used, unfortunately abridged to TW by the specialists who use it, which increases the confusion. To give an idea of what we mean, a TW-year is approximately equivalent to a billion tons of coal equivalent (10^9 TCE). (Throughout this book, the tons referred to are metric tons (1000 kg).)

The TCE, the TOE and the kW.h (or TW.h) are units of energy, whereas the watt and its multiples are units of power.

According to the official Oil Committee, 1 TOE is equivalent to 1.5 TCE and 1000 kW.h are equivalent to 0.333 TCE.

The UN adopts another figure with 1000 kW.h equivalent to 0.123 TCE. This difference in value arises from considering on the one hand equivalence in terms of production and on the other hand equivalence in terms of consumption. This question considerably complicates all energy balances.

LIST OF TABLES

LIST OF FIGURES

INTRODUCTION:
FROM TERAWATTS TO PSYCHODRAMA

> We have arrived at the point where we can no longer put up either with our vices or the remedies which would cure them.
>
> LIVY

After a quarter century of activity and reflection related to energy problems there was a time when I dreamed of following in Rousseau's footsteps and writing the *Confessions of a Lonely Analyst*. These confessions could be summed up in a clear-cut admission of humility: when one has said that one does not understand a great deal about future energy demands and that one does not know much more about the resources called upon to satisfy them, one has said almost everything.

So do not expect to find here *the* new solution, or even to re-encounter one or another of the solutions certain people have proposed or are attempting to impose. This book simply presents some reflections on New Sources of Energy and, with respect to these, some reflections on energy *and* society.

In this area, as in others, the time of simple schemes is past. The needle of the world's energy compass is today swinging wildly and no longer firmly points towards a continuous increase in consumption with successive substitution of more and more sophisticated sources. We have not yet got over the magnetic storm — the oil crisis of 1973-74 — that we had to go through, although, counting heads, we are waking up to the idea that we have come out of it alive.

Faced with the future and its multiple possibilities, what makes the past seem so simple is its uniqueness. One profits from this by deriving deterministic laws and extrapolating them into the future. As regards energy, this method seemingly used to work rather well. Just think of the "law" of the doubling of electricity consumption every ten years. From this apparent simplicity arose the leitmotiv of growth: growth of needs, growth of consumption, growth of available resources by playing with substitution (from one form of energy to another), growth in the size of plants and networks, and even growth of rates of growth, leading first to a besotted bewonderment and then to vertigo in the face of exponential growth (features of the first report of the Club of Rome).

FROM THE HUMAN WATT TO THE THERMONUCLEAR TERAWATT

Nothing better illustrates this double phenomenon of continuous growth, continually given new impetus by the substitution of new sources of energy, than such historical-futuristic analyses of "market penetration" against the background of the terawatt as made by C. Marchetti, who, together with Herman Kahn, is one of the stubborn proponents of growth at any cost.[1].

Let us start with the history of the terawatt. It dates back almost to the days of paradise on earth — without machines, except perhaps the angels' trumpets — or to those primitive societies that people nowadays are pleased to imagine were happy. Perhaps they were, anyway, without knowing it, not having any standard of comparison. The only machine we find is the human one, and the power available from it is measurable in watts (which haven't been invented yet!). The first "machines," water mills, then windmills, introduce the realm of the kilowatt (kW, a thousand or 10^3 watts); they settle in there; they grow there.

With the first electric power stations — this is admittedly just a schematic illustration — we reach the domain of the megawatt (MW, a million or 10^6 watts) and soon go beyond it. Passing over the intermediate powers of 10, it can be said that the most powerful machines of today — big hydroelectric dams, followed by classical thermoelectric power plants or nuclear ones — have reached or even exceeded the gigawatt (a billion or 10^9 watts) and are linked up in networks or systems with 10 to 100 times as much power.

It is only a step further from there to dream of machines and plants that are still more powerful, eventually reaching the terawatt level. Certain scientists such as Marchetti and Häfele have begun to elaborate on "the terawatt society."

Anybody who gets shivers down the spine at the idea of the terawatt — as a symbol of hypercentralization — should note that even now, prior to the arrival of machinery or plant in the terawatt range, there *already* exist systems reaching or exceeding this power level. The American electrical power system is approaching it (at about 0.8 TW). The Persian Gulf itself is almost a point source—looked at on a global scale—of "power" exceeding a TW. Big nuclear parks currently under discussion might have an average power of 50 to 100 GW, i.e., 0.05 to 0.1 TW. The idea of super machines or super power stations on the order of a terawatt is only the logical conclusion of an anticipatory step taken by extrapolation, in a process begun thousands of years ago.

The relative reduction in specific costs as the size of an installation increases is known as "economy of scale." Unit sizes (electric power stations included) have grown more rapidly than rates of consumption; in other words, the number of these installations has relatively diminished.[2] Nevertheless, this phenomenon has its

[1] One of his most recent publications "simply" studies the case of a world population of 1,000 billion persons.

[2] In France, for instance, in 1950 there were 71 thermal power station (226 units) producing 8.9 TW.h; in 1977 there were 37 power stations (121 units) producing 96.8 TW.h. That is, there were half as many thermal power stations and ten times as much electricity.

limits, once a certain size is reached, in terms of the "diseconomy of scale" due to the increasing complexity of the machines, increasing costs for transport and distribution of the energy produced, and the vulnerability of overconcentrated networks, too dependent on a few super units.

This increase of unit size has naturally been encouraged by the growth of systems and networks, in turn made possible by the substitution of new forms of energy with better performance than old and traditional forms of energy, which have been arriving at, or approaching, certain limits, for instance in the fuel sector. In effect, fossil fuels represented the "solution" adopted to solve the problem of ever-increasing energy requirements. One may well ask today whether this solution was not a fateful one. We shall come back to this point when discussing solar energy.

Having travelled from the watt to the terawatt, we must now see how mankind has progressed from wood to thermonuclear power.

Traditional wood fuel was gradually replaced by coal,[3] modestly and locally at first and with some reticence, as people are pleased to recall nowadays. The development of coal, its production and utilization dramatically accelerated the evolution of western society: evolution became revolution, the Industrial Revolution.

Then, at the beginning of the present century in the United States, and in the middle of this century in Europe — it being all to easy to forget that the primacy of liquid hydrocarbons is less than twenty-five years old — oil, a new form of energy, was substituted for coal.

During World War II in the United States, and during the 1960s in Europe, gas, yet another new form of energy, began its fulgurating penetration.

Now, before the end of the decade that saw the penetration of natural gas, a rapid increase in energy requirements and a mad rush for progress and growth, the concerns of "national energy policy" have led to the launch of the latest new form of energy, nuclear energy, and are paving the way to further new forms: thermonuclear energy, solar energy, and possibly geothermal energy. Incidentially, it may be of interest to note that this is the first time in the course of the long history of substititions that humanity has been confronted with a choice between various solutions — a choice, that is, unless requirements are such that all these solutions will be equally necessary.

Mathematical and historical study of the waves of energy developments led Marchetti to propose "laws" relating to their rates of penetration and, pursuing the terawatt, to conjugate the future tense of these laws (See Figure 1-1).

These laws show that in general it takes between fifty and a hundred years for a new technology to gain approximately fifty percent of the "market." The technology does not necessarily relate to energy, for this relatively exhaustive study looked at techniques as diversified as steel or synthetic textile manufacture.

[3] At present many developing countries are moving directly from agricultural waste products and wood to oil, bypassing coal.

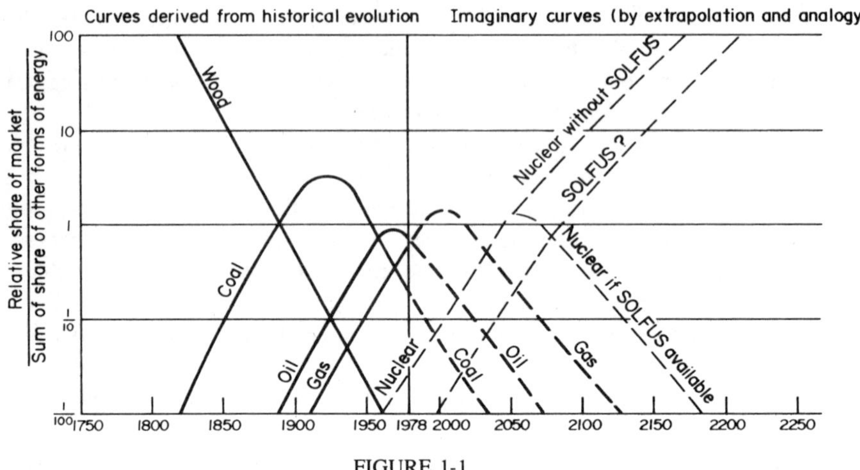

FIGURE 1-1

From these comparisons sprang the idea of stretching the historical curves found for wood (in practice a falling curve), for coal (growth, followed by a relative decline), for oil (growth, followed currently by stabilization) and for gas (growing fast) by projections showing their possible future trend, of the "growth-decline" type (or youth, maturity and old age) traced on the historical and composite pattern of evolution of just wood and coal. It should be noted that the decline of coal is questionable at present.

Curves of the same type have been constructed for nuclear power, still in its infancy, for thermonuclear and solar power, for promises, for any new future form of energy, already baptized SOLFUS (a portmanteau word from SOLar and FUSion).

The idea is appealing, if not scientifically unobjectionable, but dangerous. Its merit is to draw attention to the inertia inherent in the world of energy, to the slowness of penetration and capture of markets by any new technology, an effect that promoters of nuclear energy in the past and promoters of solar energy today tended or still tend to forget or underestimate. As such, these "laws" — which of course are not really laws — provide some interesting food for thought. The danger is that, combining thermonuclear power (or SOLFUS) and the terawatt, one might come to believe — as many energy experts did or still do — in the quasi-inevitability of our energy future centred on the continuation of growth. These futurologists, furthermore, have lent credence to this idea because it was of service to them. They had in their files THE solution: for some, fast breeder reactors using plutonium (and these are the most up-in-arms, because they think they are within sight of their goal); for some others, thermonuclear fusion; and for yet others, the sun.

SECOND THOUGHTS

Many opposing factors face these confirmed believers in the growth of energy quasi ad infinitum — dreamers whose humanitarian generosity is beyond question, whose aim is that everyone should enjoy a "vital minimum," a more or less comfortable supply of energy. Such factors include diseconomy of scale (mentioned before), concern for the environment, and geopolitical difficulties in the supply of raw materials, whether for energy or for other purposes (with a warning shot, or a foretaste, possibly being the oil crisis of 1973-74, which will be remembered in the future). Nor should a still more basic factor be forgotten: doubts and questioning (always rather paralysing) concerning the ultimate aims of our society, thought to be threatened by, among other things, the massive centralization of the terawatt, implying the policeman on the horizon. Parodying Livy, one might say we have (perhaps) arrived at the point where we can no longer put up with energy or with the remedies that would cure us of it. We are torn between intransigence and indecision, between a dangerous world and a masochistic world.

The first shock in the race to the terawatt was the prick of conscience provided by pollution, among other factors. A thunderclap was provided by Nixon's anti-pollution laws of 1970 (a Nixon shooting first and aiming afterwards): these were real laws, with their constraints and their injustices. The rallying call of antigrowth was provided by the first report of the Club of Rome, in which, it may now be said, everything was wrong except the idea behind it.[4]

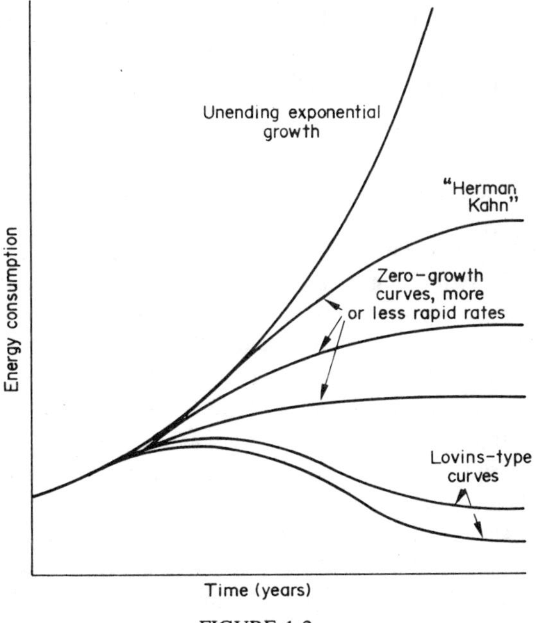

FIGURE 1-2

[4] It would be premature to deduce from these examples that only false ideas lead the world.

It is amusing to trace "geometrically" the evolution of the phenomenon (Figure 1-2.). For the convinced optimists, growth can continue quasi ad infinitum. From a basis of exponentially growing needs and seemingly inexhaustible technological resources, the windmills of energy lead you to the thermonuclear stars. A first reaction — which we dare not call a realistic one in view of the example to follow — is gently to tone down this exponential curve (without breaking it!) to make it a logistical or S-shaped curve, progressively tending towards the asymptote of a limiting value. Such a procedure was followed by the optimistic Herman Kahn (in his book "The Next 200 Years"), who put the bend in at 1976, took the American bicentenary (1776-1976) and put the asymptote's approach around 2176, using values many times higher than today's: a world population of 15 billion inhabitants (as against 4 billion today), an average income of 20,000 dollars per head (currently 1,300, but with great disparities) and a world energy consumption of some 120 billion TCE (as against 8 billion TCE today).

Below this "high" curve come all the other S-shaped curves, characteristic of what might be called "an ultimate state of zero growth," reached, however at different rates, with stabilization at a higher or lower level.

Still lower than these are the curves of "negative" growth, or "Lovins curves," where the total consumption of primary energy first reaches a maximum due to acquired impetus (in indirect homage to the concepts of the inertia of energy and market penetration), then drops to levels equal to or even below current ones, although useful energy will have remained roughly constant or even have shown a slight growth. This situation could be caused by an increased efficiency of our energy systems, possibly thanks to having recourse to "soft technologies."

It is within this envelope of curves that, like a collector, we shall accumulate all our uncertainties.

"SCIENSATIONALISM" AND ENERGY

Who took any interest in energy in the past? Essentially it was the professionals, that is, industry. Since the beginning of the century it has been primarily the oil industry, which jumped the gun as regards global futurology if only by organizing the world oil market, as with the Achnacarry Agreement in 1928. No doubt this was a dangerous game. It came to a bad end, but nevertheless almost right up to the end it remained in good shape. Governments intervened occasionally, without enthusiasm, but were not really interested in the problem. In most countries, governments have been satisfied to record a growth in oil consumption, and in many instances they have even promoted it: one needs only to recall the crisis in the European coal industry caused by a very rapid switch to oil, and it was primarily oil from the Middle East. Governmental intervention was mostly limited to questions of taxes and prices, often with more and more meddlesome controls. Price regulation of oil and gas in the United States is a good example. Governments reacted only slowly, often taking a soft line because their economies had become more and more dependent on oil over which they might lose their control. Decrees and laws regulating prices and taxes in fact simply concealed the absence of an oil policy and, indeed, the absence of any energy policy at all.

As for the consumers, the better they were "served," i.e., the better their needs were widely satisfied in economic terms — even if these needs were somewhat "stimulated" — the less they intervened. Only recently have they shown any signs of concern for balancing the satisfaction of their needs against the environmental implications.

Times have certainly changed. Nowadays everybody has something to say on the subject. Energy today is a bit like Parkinson's coffee pot.[5] Like hypocrisy for Molière's Don Juan, energy has now become a fashionable word. It is a blessing and a curse, a psychodrama. Other questions equally or more important to or representative of our society, such as national defence, town planning or foreign trade (including the arms trade), are not thought of on the same scale as is energy and do not arouse such interest in the man on the street.

All the same, this problem is indeed important and complex. As we wrote in a previous book[6] (way back in 1973): "Although the energy problem is the most simple, the easiest to present in economic terms, the easiest to define in technological terms, we think it is closely connected with the overall problem. In this respect, the response to it, which will perhaps be one of the first answers to be provided, will be of the highest significance. Let us hope that we shall put the right value on it!" In other words, we were thinking: how can we hope to resolve the other problems threatening us if we do not even know how to resolve the energy problem? Five years have passed, and the problem has not been resolved. It has swollen up, like spaghetti: pull at one thread, and the whole plateful of society's problems comes along too.

The successive waves of energy questions have spread in growing circles, like the waves from a pebble thrown into a pond. The first overflow was in purely professional circles, in the nuclear community after the war; exchanging military atoms for civilian atoms, nuclear physicists became energy experts. Their way of thinking, too mathematical and too little in contact with the human sciences, has produced some of the dangers, by no means the least, with which the atom threatens us. In the majority of contemporary energy programs one keeps coming across this excess of nuclear physicists. They put the stamp of their thinking on these programs — and their ideas of *their* solutions — thus helping to falsify the whole situation.

Then the waves spread, reaching such other areas as the press, environmental protectionists, the consumer protectionists, and so on. Doesn't everybody have his

[5] According to Parkinson, at board meetings million-dollar decisions are often taken in minutes but the decision to buy a new coffee pot may take hours of discussion.

[6] *Ce monde affamé d' énergie,* R. Laffont, 1973. This book was written *before* the 1973 crisis, but *after* the first shock provided by the Teheran negotiations, which drew attention to problems that since then have constantly increased and have not been resolved.

own ideas on how to make the best coffee?

To capture and retain the attention of a public hard to interest, the scientists and researchers became jugglers and futurologists, artisans of the "sciensationalism" that today obscures rather than enlightens debate, while stirring it up. Their opponents join in the bidding battle. The polarization for or against nuclear energy (the "anti" press is by far the most important and threatens us with a holocaust; the "pro" literature on the other hand offers us poverty and unemployment if we reject it) and the permanent appeal to public opinion has led to a certain sciensationalism, as can be seen from such titles as "Poisoned Power," "We Almost Lost Detroit," "The American People Are Living on Top of a Nuclear Volcano," "The Doomsday Book," and "Careless Atom."

The situation is rather similar in regard to the theories of the Neo-Malthusians or prophets of doom such as Meadows. Such sciensationalism is also to be found in Carroll Wilson with his Workshop on Alternative Energy Strategies (WAES) (or in the CIA), forecasting a new and dramatic oil crisis for 1981, that is, choosing from a range of possible scenarios one of the most alarming ones. The choice of the most alarming scenarios makes use of sciensationalism to ensure that public opinion is alerted or mobilized, generally in good faith; but these alarms, often contradictory in their extremism and in their arguments, mean that their authors run the risk of losing all their credibility. Reading them, one is inclined to ask oneself if life itself will be permitted in the future, between famine and "guzzling" energy, between "nucleofascism" and an impulsive form of ultrademocracy (as foreshadowed by the America of Carter's energy policy!). One might well ask if less extreme positions might ultimately lead to a certain consensus. Perhaps this is the subconscious idea that led me to write this book.

OUTLINE OF THE BOOK

In order not to give the wrong answers or a priori ones — a failing that regrettably must be attributed to a large number of energy models designed to produce the required answer and to serve as incontrovertible "evidence" — we shall first try above all *to understand the questions better,* questions that cannot always be answered but cannot be ignored either.

In one sense everything begins and ends with consumption of or the "demand" for energy, in terms of both quantity and quality. In reality, man does not ask for "some energy, please" (an intermediate good) but expresses and satisfies needs for food, shelter, comfort, mobility, communication. Corresponding to the technical status of the society are a certain number of *means* for responding to this elementary demand, and it is these means (and the associated means needed to produce them) that in general consume energy sources in a certain form, depending on the techniques used. This finally used energy is generally not usable in the form in which it is found in nature (except solar energy used for heating, one of its advantages), and before achieving its final form it has to go through a whole series of transformations. It is even possible that the form of energy finally consumed has no resemblance to the primary resource that provided it, such as oil, coal, or uranium. It is the Gordian knot of conversion, of "secondary" energy, that is developing because its forms are those best adapted (or most flexible) for the ultimate phase of utilization by the consumer.

The study of energy demand is generally concentrated on the two ends of the chain: a detailed analysis of final or useful needs and an estimation of requirements in terms of primary energy. The latter brings us to the geopolitical problem of energy supply, which depends on the possible abundance of primary resources and their geographical location. It is not easy to go through and analyse the links that make up the energy chain, because they branch at numerous points and because of technical evolution.

The most difficult thing is to estimate, on the basis of historical evolution and the pattern of energy needs today, what will be the needs of a citizen of the twenty-first century. We cannot make heads or tails of this energy problem. Of course, one is immediately tempted to put the question: why bother to estimate the needs of our children? "They're sure to find a way. . . ." The first reason that this is not entirely true is the inertia of the energy system: today it takes between eight and twelve years to build a power station. As it will be necessary to supply this power station with fuel (coal, oil or uranium, setting the sun aside for the time being) for thirty-five to forty years, it is normal, almost essential, to give some thought today to the possibilities of this supply and consider a fifty-year stretch of time.

A second reason, if one accepts the idea of successive substitution, is the slowness of substitution, already underlined, which is just another aspect of the inertia of the energy system. It is just as desirable to avoid taking society along a dead-end road, ending in penury, as it is to compel it to convert one form of energy into another until catastrophe intervenes. The ideal would be to open up a number of options, if not all, not to dismiss any, and above all not to condemn ourselves prematurely to a single one. An estimate of how the demand for energy can or might develop over time is a preliminary step toward finding possible solutions. Now, as the chapter on energy demand will show, we are far from knowing this, far from comprehending it, whether the values sought are obtained by the simple rule of three weighted with a little common sense or whether highly complex models are used.

Today or tomorrow's level of energy consumption (presently known fuel reserves being used up in the meantime), possibly seasoned with the desire for national independence as regards energy, subtends the need to convert to so-called new forms of energy or at least to some different from the currently dominant forms threatened by exhaustion of supply or unsatisfactory for geopolitical or economic reasons. What are today's "new" forms of energy? Essentially, there are four (or five, if coal is included because of the importance of coal resources, abundant and potentially utilizable in new forms, liquid or gas). These four candidates are nuclear energy from fission, nuclear energy from fusion, solar energy, and geothermal energy.

1. Energy from nuclear fission. This is a disturbing form of energy because of its immense power and the threats it entails, ,but it looks as if it might be available tomorrow. Is nuclear fission, as planned and decided upon some twenty to thirty years ago in conditions vastly different from those of today, really the long-term solution to the energy problem, as was once believed? Will Carter succeed in imposing a new nuclear policy that will mean that Europe will lose some of the key positions acquired with so much difficulty? Is nuclear non proliferation[7] really feasible, or is it just a question of delaying tactics, like the Non-Proliferation Treaty? A big nuclear dilemma is the amount of effort involved. The several hundred billion dollars already invested worldwide tends to mean sticking to existing solutions, while there is a need to look for better ones. Currently the world is concentrating more and more on the least effective kind of reactor conceivable just because it happens to work and possesses the merit of existing. Whether it is a ''good'' thing or not, the developing countries want to get into the nuclear act, although generally nuclear power is least suited to their needs, fearing an irremediable break between the nuclear North and the nonnuclear South. Because of these contradictions and the opposition they arouse, at the end of this nuclear chapter we are led to put two questions: will nuclear development come to a halt? and *must* nuclear development be halted?
2. Thermonuclear energy (the most sophisticated form of energy envisaged today[8]). Could the technological breakthrough expected from thermonuclear energy within the next ten years change the entire basis of the nuclear problem or even of the energy problem? Contrary to what is often said, nuclear fusion is hardly any ''cleaner'' than nuclear fission. However, it would be for humanity the last step on the road to the terawatt. And certain sciensational futurologists like to imagine the civilisation of tomorrow or the day after tomorrow with a few super thermonuclear power stations buried in the Pacific, like a new Persian Gulf, whence long convoys of tankers filled with liquid hydrogen head for the five continents, each of these hydrogen tankers being potentially as dangerous as a hydrogen bomb, as was ironically remarked by Edward Teller, the father of the hydrogen bomb.
3. Solar energy. Oil is transformed into thousands of products: LPG[9], gasoline, jet fuel, gas-oil, naphtha, light and heavy fuel oils, bitumens and the whole range of petrochemical products. Nuclear power produces practically only one product: electricity (and perhaps, tomorrow, heat, as a main product or by-product). The flexibility that characterizes oil is also a feature of solar

[7] ''Non-proliferation,'' i.e., not leading to proliferation of atomic weapons across the world.

[8] Although it may be dethroned the day after tomorrow or the day after that by antimatter (Soviet communiqué, unconfirmed).

[9] Liquid Petroleum Gas.

power,[10] partly because of the diversity of possible applications and partly because of the multiplicity of its modes of domestic use, of which only a few have been investigated to date: the production of heat over an extended range of temperatures according to the degree of concentration of the incident rays; the production of electricity; the production of fuel by "energy cultures or plantations" (the gasoline bush, the hydrogen tree, or the common sugar cane for the production of methanol, a possible fuel of the future); utilization of thermal gradients in the oceans, wind power, hydraulic power, and so on. A lot remains to be done in the field of solar power, and the potential for development and progress is considerable indeed. Current research budgets are just beginning to reflect this.

4. Geothermal energy. Like Cyrano de Bergerac, twentieth-century man has multiplied the ways of going to the moon, but he has rather forgotten the richness of the earth at his feet. Geothermal energy is rather like an extreme case of oil, or more precisely, oil technology, without hydrocarbons. These are replaced by heat, the diffused heat of the earth. The basic resource is apparently immense, but the problem of domesticating it is no small one. How much of this resource could be exploited is not known at present (before the earth "quakes" at the thought). Compared with solar power, a supplementary dimension is gained — the vertical one — as well as the "burial" of operations, which has advantages and disadvantages. In the allocation of R & D budgets geothermy has in the past done no better than solar research.

The debate on new forms of energy today tends to simplify into a solar-nuclear dichotomy. But how can one honestly compare (scientific honesty being perhaps the least shared characteristic in the world) two types of energy so totally different as solar and nuclear energy, at two such dissimilar stages of development, and with such inequal treatment meted out to them. Nevertheless, this problem is doubtless one of the crucial energy problems, perhaps one of the most crucial problems of our civilization, with all its political, social, economic and ecological connotations:

- centralization of energy production to an extreme degree; concentrated at certain points on the globe, or decentralization?
- Gigantic installations (on a terawatt scale) or "human" dimensions?
- ultrasophistication (thermonuclear) or "agricultural-type machines"?
- "all-electric" (or "all-hydrogen") or diversification of secondary forms of energy, electricity, hydrogen, direct heat, methanol, and the like?

But above all, are these questions sensible, or rather, are they still sensible? In other words, is not the die already cast[11]? We shall try to answer this important

[10] There is more than one similarity between solar energy and oil. Because of their resemblance and their complementary features it is feasible to imagine a marriage of reason between them, the two providing mutual support.

[11] Because of commitment to nuclear programs, among other factors.

question. If the die is not already cast, shall we not find ourselves cornered between the urgency of the problems and the fear of too-hasty decisions?

The key to a certain "urgent need to wait" is in the future (unforeseeable) evolution of energy demand, and at the same time in estimates and availability, in terms of quantity rather than price, of the classical fossil resources, coal, oil and gas, and the possible "cousins" of these, the unconventional forms of the latter two heavy oils, tar sands, oil shales, gas from geopressured zones or tight geological formations, and so on.

As regards coal, world resources are such that one may well ask whether after its brief infidelity, an affair with oil, our civilization will not return to its domestic hearth of coal. And whether King Coal will reign over his old realms, or whether he will be ornamented with new allures (liquid or gas). Where coal is most abundant — not in Europe, alas — it is today much cheaper than oil at world market prices. (However, oil remains supreme as regards its minimum cost of production on the Persian Gulf and elsewhere.)

As for the hydrocarbons, beyond *assured reserves* (thirty years at current consumption rates for oil, and more for gas) there are *resources* still to be discovered and/or produced, possibly with new techniques (tertiary or enhanced recovery for oil) that will become progressively more attractive as OPEC prices increase. The quantity of such resources and the possible timetable for their exploitation can change the basic geopolitical energy situation: perhaps this could already be seen (as we were one of the first to predict) with the oil "surplus," which was a source of embarrassment to OPEC, negating the alarmist warnings of the CIA and Carroll Wilson, among others. One might also note the enormous trumps in oil held by the American hemisphere, which could influence the North-South debate: Canada with Arctic oil and gas and tar sands; the United States with their dominant position in technology and still with immense resources (by no means negligible supplies of oil and gas, and above all, oil shales); Mexico, with increasing reserves; Venezuela, with heavy oils in the Orinoco Basin; the whole of the Caribbean basin; and also the South American potential. Another important consequence: if there is more oil than is assumed, or admitted to, this would allow time to be bought, verifying that time engenders time. For time gained before leaving the era of oil would in fact enable this era to be prolonged thanks to nonconventional forms of oil and gas, more diluted or more difficult to produce or use, and thus more expensive; but of these there are apparently enormous resources.

What should be done with the time so gained? Essentially, it should be used to arrive at an objective comparison of long-term solutions, especially solar and nuclear power. It may be objected that this is perhaps conceivable on a worldwide scale or for certain large countries well endowed with fossil fuels — certainly led by the United States or the Soviet Union — but that Europe, especially southern Europe, is poor in energy supplies, in coal, in oil, in gas — and it is hardly less poor in uranium. The possibility to choose and the time to choose in would be a luxury for the rich. But what audit office could put a figure on the cost of too-hasty decisions?

In our current economic and social systems we do not know what criteria to use or what methods to apply in balancing "apples and oranges" — the nuclear apple and the solar orange. The former is a fruit of dissension. The latter appears distant, even Utopian. This book and these reflections of an independent "energetician" are aimed to help the reader understand these better in the search for a choice based on reason.

1

ENERGY DEMAND — AN UNKNOWN QUANTITY

In the beginning is demand . . . .

It is normal, and practical, to tackle the study of future energy problems by looking at demand,[1] because it is this that determines the choice of energy "systems" (the complex ensemble of links between resources in the ground and final consumption), together with the timetable for investment. How much energy shall we need in ten years, in twenty years, in fifty years? Where shall we all be living in half a century's time? Will there be a shift to an "oecumenopolis" with 90 percent of the world population on ocean shores and river banks, living on a ribbon development like the French Riviera? Or will the move be to nuclear towns like Houston, all center and no circumference? What will be the role of the motor car, and what kind of car will it be? How shall we live? How many of us will there be on Earth?

The estimates of world energy demand put forward by the United Nations (Léontief) or the World Energy Conference, the International Institute for Applied Systems Analysis in Laxenburg, Austria, the Club of Rome, Alvin Weinberg, Herman Kahn, and others vary by a factor of two or three for the year 2000 — tomorrow — that is, between 10 and 30 billion tons of coal equivalent per year, or by a factor of 10 to 20 for the horizon of 2020-2030 — the day after tomorrow! — that is, between 15 and 300 billion TCE per year. Although the variation factors are a little less broad — say, between two and five — the situation is hardly more satisfactory when one passes from the globe to a single country.

In order to understand why there are such variations, and perhaps to try to arrive at some ideas of our own, it is interesting briefly to analyse the forecasting tools (deductions derived from the past) or prospects (the imagined future) for energy. The choice of methods is not without hidden motives; the choice of hypotheses is often clearer still. Both reveal, implicitly or explicitly, a certain vision of the

[1] Because the supply and demand for energy are generally in balance, demand is ultimately equal to consumption. As regards the future, it has become customary to speak of demand (forecast of actual future consumption).

future, tempered both by ignorance and by hope in varying proportions. And both invite us to indulge in serious thought on the society of tomorrow.

THE CRYSTAL BALL, THE RULE OF THREE, AND SCENARIOS

The two "ends" of energy demand are primary energy and useful energy. Governments are generally more interested in the balance of primary sources of energy — coal, oil, gas, uranium, hydroelectricity, geothermy, incidental solar energy — the more so because their impact on the national economy is significant (for example, "the oil bill"). Analysis downstream goes ahead by leaps and bounds towards "sectors of consumption." The individual, on the other hand, is more interested in useful energy, as needed for heating, kilometres covered, and the like. In fact he is buying *not* useful energy, but final energy[2], fuel or electricity, gasoline etc., which is transformed into useful energy by more or less efficient devices such as boilers, radiators, or engines. Between these two ends, that is, between useful energy downstream and primary energy upstream, there is a whole range or flux of complex and interconnected links of transformation and distribution, along which it is necessary to proceed when one commences the analysis at the useful energy end (proceeding upstream very carefully), while the study of primary energy can in general dispense with this.

It must be understood that the methods and anlytical tools used are completely different, depending on whether one attacks primary sources or useful energy. It is the latter analysis, the more recently tackled, which is more promising and more valuable, because it observes and dissects the mechanisms of our society. It lends itself better to socioeconomic scenarios, which themselves may vary widely: to extrapolated scenarios, that is to say, accepting a relative continuity of structures, or to anticipatory scenarios, not hesitating ultimately to question everything, which naturally results in extremely contrasting results, often highly controversial but nevertheless interesting. The theories put forward try to explain the past, to justify the present, or to forecast the future. Often they are based on mathematical models, simplified images of complex reality. But let us never forget that a "model" is not, and cannot be, a "decision-maker." It is just a tool for exploring — with great caution — the consequences of certain hypotheses.

The simplest methods or models generally used for world consumption, for example, for the consumption of primary energy or fuel and in rare cases for consumption in certain sectors or industrial fields, take time or space as a "descriptive variable."

Time

Here we are essentially concerned with historical statistical series concerning the consumption of energy over time, neglecting other factors. Using mathemati-

[2] There is a simple schema: primary energy → (possible secondary energy) → final energy → useful energy.

cal regression techniques, one looks for the best fit of the points obtained to a curve (for example, using the method of least squares), which may be a straight line, a parabola, an exponential curve, a logistic or S-shaped curve; and one projects this curve into the future. As the choice of curve is not rigorous, the extrapolations' uncertainty (measured by the "confidence interval") grows, the further forward they look. Taking the year 2000, for example, the difference between the two most extreme estimates may represent a factor of two or three. These methods are of most interest when there are long periods of stability, which is evidently no more the case nowadays. The most famous example led to the "law of doubling every decade," once thought to apply to electricity consumption.

Space

Another method consists of comparing the evolution of a country's energy consumption with that of another country that is more "advanced" and supposing that these two countries will continue to evolve in parallel fashion, but with a certain time-lag. This entails imitating some type of idealized civilization to produce a curve, whether the imitation is deliberate or not. Thus western European countries have generally "followed" the United States[3] and in turn have been chosen — or have offered themselves — as examples to be followed by the developing countries. There is no need to go into great detail as to how questionable this method of spatial transposition is nowadays. However, this should not be taken to imply that it should be easy to find other "types" of development. Just think of the interest and also the reluctance engendered by the "Chinese example," the idea of perpetual recycling.

Attempts have been made to find more satisfactory descriptive variables than time or space, for instance, in the economic field. The two variables most used in so-called econometric methods are GNP (Gross National Product) and the price of energy, which is rather difficult to define on a national aggregate level (what is the price of an American calorie?). One possibility is to define the important relationship between consumption of energy per head of population with the revenue or GNP per head, or to define "elasticities," representing not absolute values but the growth of energy consumption relative to unit increases in GNP or prices.[4] Energy elasticity related to GNP is generally positive, with consumption increasing with the GNP, while energy elasticity related to price has in theory (and currently) negative elasticity. That is to say, the consumption of energy declines when prices increase (though this is not always easy to verify over a long period).

Taking "historical" elasticities, attempts are made to deduce future consumption by making hypotheses about the future growth rate of the GNP (back to the old problem of growth rates) or the future price of energy (about which unfortunately

[3] It is often said of some American innovation that it takes a decade to "cross the Atlantic."

[4] If, for example, when GNP increases by one unit the consumption of energy also increases by one unit, the elasticity is equal to 1; if energy increases by only 0.9 units, the elasticity is 0.9.

nothing is known). Such is the precariousness of forecasting methods. They entail the replacement of one unknown, energy consumption, with another, thought to be better known, GNP, for which more research data are available. On the other hand, analyses of price elasticity are a useful tool for tackling energy policy by fiscal means.

Two remarks may be made about these widely used (in the absence of other) methods. Elasticity varies over time. In the United States, for example, the average elasticity of energy with respect to the GNP was 1.4 between 1909 and 1920, 0.75 between 1922 and 1940, and 0.8 after the war. As a general rule elasticity breaks down in a period of crisis or discontinuity, which is precisely what we are living through now. Nowadays the belief that GNP and energy consumption are ineluctably linked is being increasingly called into question. It seems possible to conceive of types of economic development in which growth of GNP would not be systematically accompanied by as big an increase in the consumption of energy as in the past.

The second remark is an invitation to caution with regard to methods using "regression" or statistical correlations. In a very serious and very interesting study, Jacqueline Fourastié looked at the French Gross Domestic Product from 1951 to 1957, derived forecasting models, and applied these to the period from 1957 to 1966. These models used two types of descriptive variables: the classical, "serious" ones: capital stocks, added value, number of working hours involved (all traditional variables accepted by econometrists) and purely arbitrary and rather bizarre variables: the years of the calendar, the temperature in Paris, and the age of the president. The conclusion was that, viewed purely from the point of view of statistical correlation, the best model was that which simply used the weather as an explanatory variable. Similarly, the function involving weather, the temperature in Paris, and the age of the president gave a better fit than the classical function involving labour and capital, from which one may conclude that one should not be too systematic about looking for the best correlation, because it may be a fortuitous one.

AT THE OTHER END OF THE ENERGY CHAIN: THE CONSUMER

If one abandons the macroeconomy to take a look at individuals, the simplest step to take is obviously to attribute to each "consumer" an arbitrary but reasonable energy ration.[5] (The method is not normative, that is, it does not provide a general description of the paths followed to achieve the goal.)

If this value per head is multiplied by the presumed size of the population, one obtains a future value for national consumption, regional consumption (Europe,

[5] This ration is sometimes based on wishful thinking, for example, if one supposes that in twenty years an Indian would reach the energy consumption level of a European today, or if average per capita consumption, currently about 2 TCE p.a. would by the end of the century reach the current European (6 TCE p.a.) or U.S. (13 TCE p.a.) level.

Africa, and so on), or world consumption. Such simplistic calculations, quoted only as a reminder, lead to value envelopes, which are unexplained, unexplainable, and explain nothing but which keep cropping up with much more elaborate values. The most promising approach, as stated before, looks at useful or final energy, related to the individual or to homogeneous "modules" carefully distinguished and defined: for example the heating needs of a household of a particular social category in a town with under a million population, or the rail transport requirements of people travelling for professional purposes. Such defined groups — as proposed by the Institut Economique et Juridique de l'Energie (the Economic and Juridical Institute for Energy) in Grenoble,[6] France, for example, splits up the socioeconomic system into a certain number of modules. (The difficulty of manipulating these, of course, increases with the number of modules.) The energy requirements of each of these modules are analysed in detail — which implies an impressive and extensive statistical apparatus — in terms of sociocultural determinants (ambient temperature, size of home, and so on) or technical determinants (heating equipment, insulation, and the like) and other indirect determinants related to prices, GNP, housing policy, and so on. Taking these requirements based on modules, in order to foresee the future or, as it is sometimes put, "to define the parameters," certain scenarios are developed. The care needed to define the determinants, direct and indirect, leads to a preference for scenarios involving extrapolation, that is, projecting in one way or another features from the past rather than introducing discontinuities or too-radical changes, where parameters would be difficult to define.

Studies of this type are not as rare as one might think. But they are often limited to specific sectors, where they are one of the prime tools of the analysts of energy conservation.

A less detailed form of analysis involves a definition of "reasonable" requirements, generally identical to current individual requirements, but with these perhaps being satisfied by more efficient means, that is, by means using less energy. These come close to the class of scenarios or normative schemas of anticipation currently in vogue, which take other types of growth (A. Lovins) or other types of society. According to the case being studied and the state of development of the study there may or may not be a scenario illustrating the transition: the task is by no means an easy one.

MOTIVATIONS

In forecasts of future energy demand, it is important to distinguish between completely independent studies, where demand itself is studied for the sake of scientific comprehension, and captive studies forming part of a larger-scale project, with avowed aims that are more or less explicit. In other words, the choice

[6] The "Médée" model of Bertrand Chateau and Bruno Lapillonne.

of models is not always "innocent"; and the choice of extrapolation hypotheses is even less so. This is not always realized if the "promoter" is of some stature.

This comment is equally applicable to two tendencies, that of "finding" or justifying relatively high levels of consumption and that of finding or justifying relatively low ones.

The tendency towards high values can perhaps be illustrated by a historically interesting case, for, in a certain sense, it started the bandwagon rolling. In 1971, at the Fourth International Atomic Conference in Geneva, Alvin Weinberg conjured up the possibility of a world with a population of 15 billion inhabitants in the second half of the twenty-first century (at the time, not so unreasonable, even if disquieting), with each inhabitant having at his disposal on average double the current annual American "energy budget," equivalent to consumption of about 25 TCE (tons coal equivalent) per head. It should also be noted that this level was partly caused by the size of the population itself, which would induce new energy requirements such as for intensive agriculture, large-scale desalination of seawater, and the like. It also took into account some increasing inefficiency across the whole system (an idea developed independently by Ivan Illich). Ultimately world consumption would level out at 375 billion TCE annually, about 40 to 50 times the current level! At such a level of consumption, world coal resources — known and assumed, recoverable or not! — would last for about 30 years, and world oil resources — known and hoped for — between one and two years. The only solution at the time seemed to be nuclear energy, a somewhat frightening prospect, even to Weinberg, with 24,000 reactors of 5000 MW (e)[7] each (five times as powerful as present-day reactor), these being constructed at a rate of two per day over the world (to maintain the number in operation, because of obsolescence after 30 years).

This study played a pioneering role in the energy world initially, to be followed by rejection. Subsequently many studies have been inspired by it, and many currently try to distance themselves from it as much as possible. The former slowly descended the ladder of individual or global consumption; the latter voluntarily descended four rungs at a time.

Although this is not an absolute rule, it can be said that generally the highest forecasts of energy demand, whether on a global or on a national level, have nuclear power in mind or seek to justify it.

Examples are the research done by the former U.S. Energy Commission, the U.S. Energy Research and Development Administration, the International Atomic Energy Agency, and the World Energy Conference.

These studies usually use time-dependent[8] or econometric models. It is not uninteresting to note that these studies deal with today's situation, including the persistent recession, only with considerable reticence and are periodically and

[7] Megawatts (electrical).

[8] A series of annual statistics, as described above.

regretfully revised downwards. Of course, those who gladly accept nuclear power or seek to justify it also adopt with conviction the lowest values put forward for oil resources and readily doubt the possibility of a renaissance of coal. These forecasters do not hesitate to threaten us with "penury, if . . . ." accompanied by unemployment, recession, and the like (which we have already, without an energy shortage). In reality, the eyes of these forecasters are perhaps bigger than the stomach of this world starved of energy.

Opposing these forecasts are low estimates of demand, sometimes very low indeed. These studies generally take as their point of departure demand values close to the current ones for useful energy or finial energy but question the possible schemes of supply, which radically changes the balance of prime energy sources.

Amory Lovins and others such as Gerald Leach in the United Kingdom have made themselves the spokesmen for "soft" and decentralized technology, reducing electricity to the minimum minimorum (for lighting and household needs and for fixed motive forces in certain cases). As for those who think it would be impossible to realize a society in which electricity consumption would be reduced to half today's level, Lovins points out that such a state corresponds to one that existed just ten years ago.

For the time being these two types of approach are difficult to reconcile, and playing round with estimates of future energy demand and systems for meeting it sometimes looks like a chicken-and-egg problem: should the system be designed to satisfy a certain demand or should future demand be calculated to justify a certain system?

FISHING FOR RESULTS

So many studies, models, and scenarios have been presented since the oil crisis of 1973-74, that it is difficult to remember them all. The 1973-74 oil bomb has left a mushrooming fall-out of papers and reports. Everybody can fish around for the figures that suit one best.

A common element is that the majority of these studies uses the econometric method, that is, extrapolating the past (although both history and science do not repeat themselves), dicing firstly with the growth of GNP, varying from 1 to 6 or even 7 percent per annum, according to the case and the country in question, with a gradual diminution over time (taking the periods 1975-1985, 1985-2000, 2000-2020, etc.) in implicit recognition of the fact that growth cannot continue indefinitely, and dicing secondly with the elasticity of energy in relation to GNP. It cannot be repeated too frequently how subjective is the choice of values during a period of such uncertainty as we are now experiencing, this choice all too often representing wishful thinking on the part of the authors. Incidentally, these often belong to the same family, for instance, the clan of government planners, or the family of oil companies, who are very active in this area (as their future is at stake!). These families often have similar patterns of thought. To recognize this is tantamount to saying that the studies are not completely disinterested. Their seeming "convergence" is not necessarily comforting, and studies deriving average values from them contribute little to the debate!

Such econometric studies have been made for the entire world, for its major geographical, economic, or political regions, and for most countries. The horizon varies between 1985 (when nearly all the cards are down, and there is not too much risk of being wrong, so there is quite a good convergence of results) and 2020-2030, most frequently stopping at 2000, date when there are already appreciable divergences. This growth of divergences is due to the well-known exponential effect, which more and more accentuates small initial differences in growth rates. On the other hand we know of no world or regional energy studies taking final or useful energy as their point of departure, though there are a few such national studies, as for the United States and for France.

As examples, we shall give some figures for the world. Table 1-1 shows detailed statistics for a large number of countries, collected by the World Bank. Among other things, it shows the big difference in consumption per head between industrialized countries and developing countries. Even the most ambitious scenarios, or the most humanitarian ones, do not bridge these differences.

The World in 2020

We shall use the study of the Conservation Commission of the World Energy Conference (Istanbul, September 1977). This study is interesting because it covers the entire world and the far horizon of 2020. Most studies generally exclude the countries with centrally planned economies.[9] The few comments that follow will illustrate the nature of the hypotheses and their reasoning.

Total world energy consumption in 1975 was 6265 TOE (tons of oil equivalent), of which 590 million TOE was wood for burning. If the trend of the period 1960-1975 continues, with vigorous economic growth and increasing energy consumption up to 2020, world consumption will by then have reached a level of 44,300 TOE, seven times as much as in 1975.[10] If, on the other hand, the extrapolation is based on the trend over the past fifty years (including the 1929 crisis and World War II), world consumption in 2020 might reach a level of 28,200 TOE (5.4 times as much as in 1975), appreciably less than in the previous hypothesis.

The experts at the World Energy Conference in fact took two hypotheses for the average growth rate of the GNP: a so-called low figure of 3 percent p.a. and a so-called high figure of 4.2 percent p.a. Taking the lower rate, income per head would about double between now and 2020; at a rate of 4.2 percent, it would

[9] Because of the paucity of data available.

[10] Given nearly double the population, this would imply that the average level of consumption per head would be multiplied by almost 3.5 (from 1.5 TOE p.a. to 5.25 TOE p.a., somewhat above the current European level).

almost triple.

These hypotheses concerning the growth rate were supplemented by hypotheses on the elasticity of energy consumption related to price, or by possible constraints on the availability of oil in particular and energy in general.

According to the hypotheses made, demand may vary by a factor slightly over two between the two extreme values (19,000 and 41,500 TOE) or between 3 and 7 times the value of 1975. As regards prices, the experts forecast they will double (in real terms) between 1975 and 2020. An important result is that even with vigorous efforts to conserve energy, the demand for primary energy would grow significantly up to 2020, reaching a level from three to nearly four times the current demand, according to whether economic growth is weak or relatively strong (and it would be nearly seven times as high without energy-saving measures).

As regards supply, put briefly, it is estimated that oil consumption will continue to grow up to 1985-1995 and then will be limited by the decreasing availability of oil, which will be reserved more and more to noble uses, petrochemical products, and especially transport (with electricity even satisfying some of the growing demand in the transport sector). Coal will respond to the increased demand for fossil fuels and will see its production multiplied by 4 or 5 (equivalent to 10-12 billion tons p.a.). Natural gas will go the same way as oil. Electricity's share of final energy will double, moving from 10 percent to 20 percent, increasing its share of the primary balance from 25 percent to nearly 40 percent, thanks to a major contribution from nuclear energy (65 percent of electricity will be of nuclear origin by 2000). The share of renewable energy would remain constant at 15 percent (hydroelectricity, wood, solar energy), and the *absolute* share of solar energy would rise from practically zero to a value approaching that of all electric energy produced today.

All in all, this is already a relatively classical scenario.

THE ASCENT OF THE ENERGY CHAIN

The concept of useful and/or final energy is very interesting, but for the sake of completeness it is necessary to ascend the energy chains up to primary sources of energy. With traditional forms of energy there were not many of these chains, and they were relatively well identified. Some are "exclusive," such as the fuel chain for cars or aircraft. Others compete with each other. For example, in domestic heating, there is the direct chain: oil → fuel oil → boiler, or the indirect chain: coal (or oil or uranium) → electricity → electric heating. The debate is well known, and the proponents of electricity, not without reason, defend the cleanness and efficiency of electricity, which almost compensate for its poor efficiency — between 30 and 40 percent of yield — at the level of production.

As regards tomorrow's energy, the problem is more complicated, and the debate grows more exacerbated. This is for two reasons: the number of chains in competition for certain applications is increasing, on the one hand; but, on the

TABLE 1-1: World Energy Production And Consumption Indicators

	Average Annual Energy Growth Rates (percent)				Per Capita Energy Consumption (kilograms of coal equivalent)		Energy Consumption Per Dollar GDP (kilograms of coal equivalent)		Energy Imports as a Percentage of Merchandise Export Earnings	
	Production		Consumption							
	1960-74[a]	1974-76	1960-74[a]	1974-76	1960	1976	1960	1976	1960	1976
Low Income Countries (g)	**6.8**	**6.3**	**5.7**	**4.6**	**113**	**166**	**0.9**	**1.1**	**9**	**19**
1 Bhutan	..	..	..	..	..	..	..	..	..	..
2 Cambodia	..	..	−0.1	(.)	31	16	..	..	9	..
3 Bangladesh	..	5.4	..	9.7	..	33	..	0.3	..	29
4 Lao PDR	..	−1.0	13.4	(.)	18	61	..	..	..	..
5 Ethiopia	14.1	2.0	14.7	−10.3	8	27	0.1	0.3	11	27
6 Mali	..	19.7	5.5	5.4	15	27	0.2	0.3	13	25
7 Nepal	27.2	8.2	12.3	1.3	5	11	(.)	0.1	..	..
8 Somalia	..	..	7.4	10.1	19	47	0.2	0.4	4	13
9 Burundi	..	19.6	..	0.3	..	12	..	0.1	..	..
10 Chad	..	..	7.2	10.5	10	23	0.1	0.2	23	27
11 Rwanda	..	3.0	..	11.3	..	17	..	0.2	..	11
12 Upper Volta	..	..	6.5	5.1	5	18	0.1	0.2	..	19
13 Zaire	3.0	54.4	4.3	−6.4	87	62	0.9	0.7	3	16
14 Burma	4.8	8.2	3.6	0.5	55	49	0.5	0.4	4	12
15 Malawi	..	17.5	..	9.8	..	56	..	0.4	..	18
16 India	4.4	9.8	4.9	7.0	142	218	1.3	1.6	11	26
17 Mozambique	3.2	32.1	5.7	0.8	114	133	0.4	0.5	11	28
18 Niger	..	..	14.3	8.4	5	35	(.)	0.2	..	..
19 Viet Nam	..	..	..	..	..	124	..	..	..	..
20 Afghanistan	39.7	−6.0	9.4	2.9	15	41	0.2	0.4	12	12
21 Pakistan	10.0	1.1	5.9	1.0	61	181	1.3	1.2	..	..
22 Sierre Leone	..	..	10.3	−3.5	31	112	0.3	0.5	11	10
23 Tanzania	10.6	29.6	10.4	12.9	41	68	0.3	0.4	..	22
24 Benin	..	..	8.8	−11.2	39	49	0.2	0.3	..	43
25 Sri Lanka	10.4	0.7	6.2	−1.9	107	106	0.7	0.5	8	24
26 Guinea	*16.1*	(.)	3.2	1.6	65	93	0.3	0.4	7	..
27 Haiti	..	18.6	2.8	4.3	36	28	0.2	0.1	..	14
28 Lesotho	..	..	..	..	..	..	..	..	..	..
29 Madagascar	6.8	4.3	8.9	4.4	38	66	0.2	0.3	9	22
30 Central African Emp.	14.2	2.0	7.4	7.4	37	41	0.1	0.2	12	1
31 Kenya	9.3	19.6	4.2	1.2	143	152	0.8	0.6	18	54
32 Mauritania	..	..	16.8	2.9	18	102	0.1	0.4	39	6
33 Uganda	5.2	−6.3	9.5	−6.0	30	48	0.1	0.2	5	4
34 Sudan	..	20.2	13.2	−2.7	52	143	0.2	0.6	8	26
35 Angola	35.3	−26.6	8.8	−6.6	86	166	0.2	0.5	6	2
36 Indonesia	8.5	6.1	4.2	22.1	129	218	0.8	0.9	3	5
37 Togo	..	..	12.5	11.9	23	85	0.1	0.3	10	19
Middle Income Countries (g)	**7.6**	**0.8**	**7.6**	**5.2**	**393**	**916**	**0.7**	**1.2**	**10**	**22**
38 Egypt	9.8	45.1	2.7	20.3	298	473	1.7	1.8	12	15
39 Cameroon	1.1	6.4	4.0	7.6	55	98	0.2	0.3	7	10
40 Yemen, PDR	..	..	−13.6	24.4	299	324	..	1.6	..	..
41 Ghana	..	1.2	6.6	−4.1	106	157	0.2	0.3	7	18
42 Honduras	29.5	9.8	8.9	1.4	155	264	0.5	0.7	10	12
43 Liberia	31.8	2.0	19.3	−5.1	86	418	0.2	1.0	3	12
44 Nigeria	37.4	−3.5	10.2	5.8	34	94	0.1	0.2	..	..
45 Thailand	28.0	21.4	16.9	4.4	64	308	0.3	0.8	12	28
46 Senegal	..	..	4.6	5.5	121	156	0.3	0.4	8	15
47 Yemen Arab Rep.	..	..	12.7	38.9	7	41	..	0.3	..	..
48 Philippines	5.6	9.1	9.6	7.7	147	329	0.6	0.8	..	..
49 Zambia	..	6.0	..	10.1	..	548	..	1.3	..	5
50 Congo, People's Rep.	15.7	−9.6	5.2	−2.0	119	142	0.3	0.3	25	8
51 Papua New Guinea	..	..	..	..	51	289	0.2	0.6	..	..
52 Rhodesia	1.9	−1.0	..	−0.7	..	634	..	1.2	..	..
53 El Salvador	5.1	17.6	7.7	10.1	127	260	0.4	0.6	6	10
54 Morocco	1.9	−0.2	7.7	2.7	148	273	0.4	0.6	9	23
55 Bolivia	17.2	−1.8	7.0	12.4	147	318	0.5	0.7	4	1
56 Ivory Coast	9.7	17.1	15.5	3.9	76	380	0.2	0.4	5	10
57 Jordan	..	..	6.5	20.8	197	527	0.7	1.0	79	54
58 Colombia	3.4	−2.4	6.3	3.6	491	685	1.3	1.2	3	2
59 Paraguay	..	21.2	8.5	7.6	87	189	0.2	0.3	..	..
60 Ecuador	19.0	2.7	8.3	15.6	201	455	..	0.7	2	7
61 Guatemala	9.9	6.3	6.1	7.7	174	257	0.4	0.4	12	15
62 Korea, Rep. of	6.3	3.4	13.2	6.7	258	1,020	1.2	1.7	70	23

TABLE 1-1 (cont.)

63 Nicaragua	26.6	4.8	10.0	5.6	174	478	0.4	0.6	. .	. .
64 Dominican Rep.	4.4	−3.5	14.6	2.4	157	683	0.3	0.9	. .	24
65 Peru	3.5	2.0	6.2	5.7	445	642	0.8	0.7	. .	. .
66 Tunisia	73.4	−4.8	9.5	5.4	190	456	. .	0.6	15	23
67 Syrian Arab Rep.	86.2	25.9	9.0	18.2	321	744	0.7	1.0	16	16
68 Malaysia	37.4	41.3	11.1	0.4	242	602	0.6	0.7	2	9
69 Algeria	11.7	7.4	12.2	21.4	252	729	0.3	0.8	14	2
70 Turkey	7.6	4.2	9.9	12.8	245	743	0.5	0.8	16	58
71 Mexico	6.0	11.5	7.7	1.5	770	1,227	0.9	1.0	3	10
72 Jamaica	−0.7	9.0	11.2	4.2	426	1,937	0.2	1.5	11	34
73 Lebanon	12.7	−1.2	6.3	−26.0	548	533	. .	. .	68	4
74 Chile	4.0	−2.9	6.1	−4.7	845	987	1.2	1.2	10	25
75 China, Rep. of	2.3	6.0	8.6	12.8	583	1,797	1.4	1.8	. .	. .
76 Panama	14.8	1.9	10.5	7.1	448	885	0.7	0.8	. .	. .
77 Costa Rica	9.5	7.8	10.4	2.1	233	488	0.4	0.5	7	14
78 South Africa	3.8	8.1	. .	8.1	. .	2,985	. .	. .	9	. .
79 Brazil	8.1	6.4	8.6	7.2	332	731	0.6	0.6	21	43
80 Uruguay	3.7	−5.5	3.1	2.5	825	1,000	0.7	0.8	35	39
81 Iraq	4.9	7.9	5.9	7.5	487	727	0.7	0.5	(.)	(.)
82 Argentina	6.5	0.8	5.7	1.5	1,129	1,804	0.9	1.0	14	14
83 Portugal	4.4	−20.3	8.3	4.6	382	1,050	0.5	0.6	17	38
84 Yugoslavia	4.7	3.6	7.1	3.7	872	2,016	1.3	1.3	8	22
85 Iran	14.5	−0.9	15.6	9.8	270	1,490	0.4	0.8	. .	. .
86 Trinidad and Tobago	2.8	6.0	4.8	16.6	1,775	4,272	1.0	1.7	35	51
87 Hong Kong	. .	. .	6.8	5.6	468	1,313	0.7	0.7	5	7
88 Venezuela	1.2	−10.8	6.6	3.4	1,694	2,838	1.1	1.2	1	. .
89 Greece	14.3	20.3	13.2	6.2	460	2,250	0.5	0.9	26	48
90 Israel	41.9	−86.6	9.6	−1.4	1,270	2,541	0.7	0.7	17	28
91 Singapore	. .	. .	16.8	3.9	372	2,262	0.4	0.9	17	38
92 Spain	2.5	−4.6	8.5	4.1	756	2,399	0.6	0.8	22	59
Industrialized Countries (g)	**3.2**	**0.7**	**4.9**	**1.3**	**4,462**	**7,079**	**1.2**	**1.1**	**11**	**24**
93 Ireland	0.1	6.5	4.7	0.6	1,838	3,170	1.2	1.2	17	17
94 Italy	2.2	1.6	8.3	1.3	1,086	3,284	0.6	1.0	18	30
95 New Zealand	5.2	8.2	5.7	2.0	2,277	3,617	0.7	0.8	. .	. .
96 United Kingdom	−1.2	10.0	1.7	−1.4	4,861	5,268	1.6	1.3	14	22
97 Japan	−1.7	3.5	10.7	−0.7	1,171	3,679	0.8	0.8	18	42
98 Austria	1.5	−4.6	5.1	1.4	2,129	4,013	0.8	0.8	12	17
99 Finland	3.3	−10.0	9.1	1.6	1.529	5.177	0.5	0.9	11	25
100 Netherlands	16.2	6.2	8.7	2.5	2,504	6,224	0.7	1.0	. .	. .
101 France	−1.3	−3.1	5.8	−0.5	2,474	4.380	0.7	0.7	16	26
102 Australia	11.1	7.2	5.6	4.6	3,857	6,657	0.8	0.9	12	8
103 Belgium	−7.2	16.8	4.9	0.2	3,851	6,049	1.1	0.9	11	15
104 Denmark	−20.1	45.6	5.5	4.6	2,830	5,320	0.6	0.7	15	23
105 Germany, Fed. Rep.	−0.7	−0.7	4.5	2.0	3,695	5,922	0.9	0.8	7	16
106 Canada	8.9	−3.1	6.0	2.1	5,750	9,950	1.3	1.3	9	11
107 United States	3.5	−0.5	4.1	1.7	8,172	11,554	1.6	1.5	8	30
108 Norway	6.8	28.0	5.9	4.4	2,702	5,263	0.7	0.7	. .	. .
109 Sweden	3.6	9.3	4.9	5.4	3,572	6,046	0.7	0.7	16	18
110 Switzerland	4.2	−1.6	5.9	−2.8	1,873	3,342	0.3	0.4	10	11
Capital Surplus Oil Exporters										
111Saudi Arabia	14.1	0.6	14.4	31.4	267	1,901	. .	0.4	. .	(.)
112 Libya	29.1	13.0	17.9	26.6	251	1,589	0.1	0.3	83	1
113 Kuwait	4.6	−7.9	6.7	6.1	10,396	9,198	0.4	0.6	. .	. .
Centrally Planned Economies (g)	**4.8**	**5.3**	**4.8**	**5.2**	**1,378**	**2,047**	**2.2**	**2.0**	**. .**	**. .**
114 China, People's Rep.	4.5	5.4	3.6	4.8	683	706	3.5	2.0	. .	. .
115 Albania	10.1	6.9	12.5	17.1	302	867	1.1	1.6	. .	. .
116 Korea, Dem. Rep.	9.1	10.4	9.1	10.6	989	3,072	3.8	4.9	. .	. .
117 Mongolia	10.4	10.1	7.3	8.6	540	1,156	0.8	1.5	. .	. .
118 Cuba	20.6	−8.0	4.4	5.7	912	1,225	1.0	1.5	. .	39
119 Romania	5.8	4.5	8.0	7.5	1,342	4,036	3.8	3.0	. .	. .
120 Bulgaria	3.3	7.7	9.8	5.0	1,303	4,710	1.2	2.0	7	. .
121 Hungary	1.8	2.5	3.9	5.3	2,072	3,553	1.5	1.6	13	14
122 USSR	5.6	5.6	5.3	4.4	2,839	5,259	1.9	2.0	4	4
123 Poland	3.9	5.0	4.1	7.7	3,107	5,253	2.1	1.9	. .	. .
124 Czechoslovakia	1.3	2.7	3.1	4.2	4,741	7,397	1.9	2.1	. .	15
125 German Dem. Rep.	0.5	0.7	2.1	2.5	4,950	6,789	1.8	1.6	. .	. .

Figures in bands are summary statistics for each group of countries; (g) group average, (. .) not available; (.) less than half the unit shown; (a) figures in italics in these columns refer to 1961-74 rather than 1960-74; all growth rates are shown in real terms; low-income countries = per capita income U.S. $300 or below; middle-income countries = per capita income above $300.

Source: World Bank

other, the very principle of long chains and their structure are under attack because of their growing cost and their diminishing final energy return.

It is not uninteresting to note that for a time all new forms of energy were primarily studied with a view to electricity production (fission, fusion, solar electricity, geothermal power). Today the competition seems also to apply in the important sector of heating, which represents some 30 percent or more of energy requirements. For heating or the utilization of low-temperature heat there is a plethora of choice: in addition to the traditional forms of energy, coal for urban heating, oil, and gas, for a few years now solar and geothermal heating have also been available.

The complexity of the problem of chains is well illustrated by a comparison of electric heating (of nuclear origin, for example) and solar heating. In the case of electric heating, the chain is particularly complex, from the uranium mine to the reactor, from the reactor to the electricity user. This is complemented by a very complicated fuel cycle: the complexity and length of the chain and the sophistication of the equipment lead to high costs (both for the user and for society), but the push-button "product," clean, and always available, is an attractive one. In the case of solar heating, installaton is simple, and the chain is reduced (apparently) to its simplest expression: a collector on the user's roof. These cases are perhaps extreme, but they well illustrate the contrast between the two tendencies. To date, our economic and industrial development has generally seen the energy chains lengthening, going through more and more complex processes, with increasing degradation of yield. In fact, this was encouraged because of the low cost of fossil fuels. Thus we have passed from the direct utilization of a primary source (coal, among others) to the use of a final source (secondary, like electricity yesterday or today and hydrogen tomorrow), more convenient and better adapted, more flexible and cleaner in use. The necessary conversion is effected in big centralized plants, where gigantism is attractive because it offers benefits from effects of scale.

A limiting case has been made evident with the possibility of finally using hydrogen. This is in fact not a new energy *source* but a new energy *carrier*, comparable to electricity (and possibly competing with it). Hydrogen is relatively easier to transport than electricity over long or very long distances, via hydrogen pipelines or a seagoing hydrogen tanker. It can also be made from an inexhaustible raw material, water. Its combustion simply produces water vapour. So one might have imagined at a time when nuclear energy seemed vast and almost free (yes, there was such a time!) a whole series of transformations, one example of such a chain being: uranium → nuclear reactor → electricity → hydrogen (by electrolysis) → hydrogen transport → reconversion of hydrogen to electricity → redistribution of electricity for final utilization in the form of, say, electric heating. The global yield of energy between the uranium and that finally used by the consumer would not be over 5-6 percent, or at best still under 10 percent. Would this not be too high a price to pay, even ignoring the huge investment necessary in such chains of "hard" technology, for push-button convenience?

It is such an evolution, leading to longer and longer chains, and similar principles that are reexamined by Amory Lovins, who is systematically researching short and decentralized chains. The idea is attractive, and it has been cleverly presented. It merits attention, but much more research is necessary before it can be pronounced on. Lovins is researching not so much the shortest circuits but rather the most decentralized technologies possible (his bête noire being nuclear electricity) and small-scale technologies such as solar heating, diesel-electric generators, and coal-fired furnaces or boilers burning coal on fluidized beds.[11] It should be noted that the fuel production for this and for diesel-electric generators is of necessity on a large scale: it requires oil to be refined and generally calls for long transport and distribution circuits from the oil field (or the coal deposit) to the final user. Lovins rests his argument on cost considerations, showing for example that one kW.h from a hydroelectric mini-plant costs between 10 and 45 dollars per barrel of oil equivalent as against a little more than 100 dollars per barrel of oil equivalent for the same kW.h distributed via an interconnected network and produced by a light-water reactor (figures not always easy to verify).

These developments should be noted. But in fact they divert us quite a long way from energy requirements, strictly speaking.

LIFE STYLE AND ENERGY DEMAND[12]

To analyze final and/or useful energy on the scale of socioeconomic modules or, better still, by households or individuals, is to study or to try to define, to seize on "life styles" more or less tied to choices made by society (without forgetting that individuals are also agents of production).

It is evident what an enormous task this must be on a world scale, it being already quite considerable and adventurous just on the scale of a single country. What will American society or Indian society be like in the middle of the next century?

An enormous effort must still be made to help us better understand current and future real individual requirements, whatever form they will take. After so many energy studies engendered by the crisis, we feel rather like those American buses that before every level crossing must obey the sign "Stop, look, and listen."

It is in this spirit that we must open the dossier of new forms of energy, conscious, alas, of our ineptitude and inability to foresee the coming changes.

[11] Fluidized-bed combustion uses powdered coal mixed with powdered magnesium oxide, for example, which behaves like a fluid thanks to a current of gas (air). Combustion is "clean," because the sulphurous components attach themselves to the magnesium oxide. This promising technique is in the development stage.

[12] Pierre Fournier (in the left-wing magazine *Charlie Hebdo,* France) made the apt comment: "You do not change society without changing your life, but you do not change your life without changing society."

2

IS THERE SUCH A THING AS "GOOD" NUCLEAR POWER?

As I begin this chapter my pen takes a fresh breath, the breath of the past, for I used to be a nuclear engineer, and I used to devote half of every day of my active life to the atom, a long and generous half. I was not among the first to do so — the founders of 1955 were — but I was one of the first to follow the first, the "Geneva class," which had several years of atomic research under its belt by the time of the remarkable event of the first international conference on the peaceful uses of atomic energy. This was the first meeting in the history of mankind when scientists from all over the world could "communicate" on a planetary scale and thought to possess the lever with which Archimedes believed he could move the world.

And so I lived through one of the most exciting endeavours of the human spirit. In all our actions, what generosity there was! From all our experiments, what promises — naive, perhaps, but never with unworthy motives — of a better and richer life in the future, for everybody. Perhaps we were egotists, but we were sincere. *We believed.* It is hard to admit today that perhaps we were mistaken. At what point might we have erred, and how much?

For conditions have changed considerably, and the world has changed, too. Today disquieted opinion is putting the questions that, for whatever reasons, we did not ask ourselves some twenty-five years ago (whereby it is irrelevant whether this opinion is "manipulated" or not). If we are honest, we will ask ourselves these questions today, questions on ultimate safety, on risk acceptability, on the long-term storage of wastes, and the like.

In earlier days, we indeed did not realize the importance of these questions from the social and political points of view. Above all, we were scientists, and looked at the problems simply as scientists.

Now we are asking ourselves: Will nuclear power survive its difficulties, its internal contradictions, the attacks from outside to which it is exposed? For it is sickly, certainly, but still alive; even better, it "works." If it survives, will it be in the form of a "new" nuclear power or a slightly altered form of traditional nuclear power?

It is not possible to cover all the problems of nuclear power's future in a chapter and a half, or to add the 1001st argument to the debate on nuclear safety. We begin by saying that nuclear power is dangerous, or, more precisely, that in the past its dangers have been overcome but there is no guarantee that this will continue to be the case.

Having said this, we shall limit ourselves here to the four principal themes:

- the current state of development of nuclear energy in the world
- the problem of proliferation of nuclear armaments, perhaps the most severe threat to the future of nuclear power in the short term (and to the future of humanity in the long term)
- uranium resources, which are still less well known than oil resources (are we perhaps about to exchange the uncertain for the unknown?)
- the interest in, and possible penetration of, breeder reactors in the light of the two preceding problems

THE STATE OF NUCLEAR POWER, OR HOW REALITY LAGS BEHIND FICTION

We shall start with a quick statistical snapshot of nuclear power today, and we shall follow up with a dynamic perspective, looking at nuclear power tomorrow as seen the day before yesterday, yesterday, and today.

Ralph Nader, one of the firm opponents of nuclear power in the United States, in 1977 said that nuclear power was dead. Perhaps this will be true tomorrow, but today his statement is still false.

World electricity production by nuclear power plants in 1977 exceeded 500 billion kW.h, with the United States still well in the lead with more than half total production.

On 1 January 1978 there were 183 nuclear power plants in operation in the Western world[1] and 33 in eastern European countries, making a total of 216. The installed capacity of the 183 plants in the West was close to 100GW(e), representing an average of some 540 MW(e) per plant. This average is depressed by the first plants constructed, which were of relatively low power, while the majority of plants on order today are of about 850 to 1300 MW(e).

The United States was well in the lead, with nearly half the installed capacity, followed by Japan (10 percent of capacity installed in the West), the United Kingdom, and Germany. France came sixth, while the last on the list of 18 Western countries (ranked by capacity) was Pakistan with 0.1 GW(e), about a

[1] Note that a plant in general, and more and more frequently nowadays, is made up of several units, or nuclear reactors. Power is expressed in gigawatts or megawatts. (1 GW = 1000 MW = 1 billion watts.) To avoid confusion, often the terms GW(e) and MW(e) are used, the (e) standing for electricity.

thousandth of the installed capacity in the West.

These 183 plants on 1 January 1978 included 13 different types of reactor, a heritage of the historical battles between various designs. Well in the lead were the light-water reactors (PWRs with pressurized water and BWRs with boiling water, 65 and 52 respectively, of American design), followed by 28 units using the British Magnox design, and 8 using the French design using natural uranium, gas, and graphite. In between the United Kingdom and France are the 11 Canadian heavy-water reactors. Then came the fast breeders, also using natural uranium, plus some others devolving from the past or with an uncertain future.

If one simply considers the 18 reactor units started up in 1977 in the Western world, the proportions change: 15 light-water reactors, reflecting world-wide concentration on this type of reactor. Five of these are BWRs and 10 are PWRs. The figures indicate the advance of this latter type. One of these is a sort of breeder promoted by the indefatigable Admiral Rickover. Other types include one Canadian heavy-water reactor, one British AGR (using graphite, gas, and slightly enriched uranium), and a small experimental German fast reactor.

In addition to these 216 nuclear units in operation, on 1 January 1978 there were some 200 nuclear units in the course of construction across the world, with a total capacity of the order of 200 GW(e). If there is not an almost immediate general moratorium, these reactors, whose construction in many cases is well advanced, will probably be completed and will bring the total installed capacity to about 300 GW(e). In a quarter of a century, nuclear power capacity will have reached about three-quarters of the world's hydroelectric capacity installed over nearly a century.

If one looks at the programs currently under way, one would probably be inclined to think nuclear power is alive and kicking. What other industries show a tripling of capacity in their order books? But stepping back a little, one might be led to ask whether it doesn't have its future behind it? Two types of consideration lead us to put this question: successive downward revisions of forecasts and the impressive list of difficulties facing the nuclear industry today.

HISTORICAL EVOLUTION OF FORECASTS OF INSTALLED NUCLEAR CAPACITY

For several years there has been a competition to see who can make the lowest forecasts for the nuclear future of the Western world (as for the USSR, it maintains a serene and constant optimism about its own program in the international meetings in which it participates). This affects all countries: "They did not all die, but all were hit . . ." Indeed, if one regards the evolution of world nuclear capacity forecasts with a high followed by a low over about ten years, it rather looks like the stock exchange.

At the beginning of the 1970s, nuclear programs were in full swing. World nuclear programs had been relaunched by the spectacular number of commercial orders in the United States in 1965-67, destined to help catch up with the United Kingdom and to prepare for the commercial conquest of the world with light-water reactors. The beginning of the 1970s also saw the birth of the environmentalist wave in the United States. (The National Environmental Protection Act was a 1969 Christmas present from Richard Nixon to the American people, some of whom would be roused to opposition). But, on the whole, times were good.

Then, in 1973-74, came the oil crisis. The nuclear tides rose, and already ambitious programs, given their industrial novelty and technical difficulties, mushroomed in certain countries, including some in western Europe (particularly in France and the Federal Republic of Germany, but also in Spain, Italy, Egypt, Iran, and elsewhere).

On the other hand, some countries hit the downswing they had forecast, headed by the United States and Japan (and for the same reason: growing delays of the programs due to public opposition). Nuclear programs also shrank in the United Kingdom, which was confronted with severe economic difficulties despite unexpected oil finds.

In 1977, the whole world was struck by environmental fever. In the United States there was a collapse precipitated by the new policies of President Carter, who might well be asked what the thinking was behind his nuclear policy; in Japan, in a bad way regarding sites for its plants; in Germany, which almost ran into a de facto nuclear moratorium, and even in France, perhaps the least affected, but where programs ran into some delay, finance being difficult and sites not easy to find. Taken together, there were some impressive slides. Most strongly affected were the forecasts for 1985 and 1990, though some programs remain hopeful and continue to make more optimistic forecasts for the year 2000 or beyond.

For the OECD area, for example, the forecasts for 1985 made in 1970 were 555 GW(e), and in 1973 still 542 GW(e) (a slight drop, principally due to the United Kingdom). In 1975 the forecasts fell within a bracket of 437-484 GW(e) but in 1977 came a new jump with closed eyes, between 259 and 434 GW(e); the lowest value, a reduction of nearly 60 percent, today still appears too high.

In the United States, the forecasts for 1985 have fallen from 277-280 GW(e) at the beginning of the seventies to 205 GW(e) in 1975 and 115 GW(e) at the end of 1977. In May 1978, James Schlesinger, head of the Department of Energy and former chairman of the all-powerful Atomic Energy Commission, talked of 100 GW(e) for 1985, a reduction of some 65 percent. The program was cut down within four years to a third of its initial objective. Incidentally, this value of 100 GW(e) for 1985 is quite close to that forecast by the U.S. Atomic Energy Commission in its report to President Kennedy back in 1962.

In France the downward trend has not quite overcome the countertrend upwards: in 1970 the estimate for 1985 was 25 GW(e), in 1973 it was 32.5, in 1975 56 GW(e) (showing impetus from the Messmer program after the oil crisis) and cut back at the end of 1977 to 34 GW(e) according to the OECD/IAEA,[2] and according to EDF[3] figures, exactly 34.2 GW(e) on 1 January 1985.

[2] International Atomic Energy Agency, Vienna, Austria.

[3] Electricité de France

Beyond 1985, estimates diverge strongly, according to whether the forecaster believes in the continuation of current trends (slowing down of the economy, sustained public opposition) or a picking up in the economy and in the order books. Thus for the OECD area, the IAEA's estimates oscillate between a low of 850 GW(e) and a high of 1640 GW(e), a factor of 2, although the experts admit that currently the low value appears the more probable.

On a global scale, and taking the same hypotheses, estimates of the nuclear capacity installed vary between 278 and 368 GW(e) for 1985 (one may note the predominant role of the OECD area with 93 percent of the world total) and between 1000 and 1890 GW(e) for the year 2000. The low estimate, 1000 GW(3), is less than half the low estimate from 1975, which was 2005 GW(e).

For the United States alone, the forecasts for the year 2000 have been cut back within a few years from 1200 GW(e) to a maximum fixed by President Carter of 300 GW(e).

These figures illustrate the extent of the reduction of programs. Nevertheless, the numbers are still impressive.

Indeed, while most of the reactors planned between now and 1985 are already under construction and will probably be completed more or less rapidly,[4] the situation is much more uncertain beyond then. Will there be a picking up in the pace of nuclear programs or continuation at a slow rate, or will things gradually come to a standstill? All the hypotheses are plausible.

However, even if they are cut back, it must be realized that these programs of electronuclearization still represent for most countries the majority of the increase in electrical capacity forecast for the next few years. It should be added that the world recession, which is affecting nearly all the world's industrialized countries, has led them to revise downwards, sometimes drastically, their programs for equipment in electrical, nuclear, and fossil fuel plants — a curious joining of forces with public opposition, though the importance of this latter should not be exaggerated.

This worldwide recession, the new background to the energy situation, is still not well understood. In addition, there is questioning of conventional assumptions such as the ''inevitable'' relation between the consumption of energy and/or electricity on the one hand and GNP on the other. Together with the undeniable difficulties associated with nuclear power, these are all factors that have led to a slowing down of programs and to great caution on the part of electricity managers, beset as they are by problems of financing.

Even slowed down, even mishandled, the world nuclear program remains significant. In terms of installed capacity it is 250 GW(e) minimum for the OECD in 1985, as stated above. In terms of productivity it is 1250TW.h or more, the equivalent of 400 to 500 millions of tons of oil per year. Finally, in terms of investment, it is 100 billion dollars. Quite a lot of these billions are invested in

[4] Although a mid-1979 study forecast up to fifty possible cancellations as a consequence of the Three Mile Island, Harrisburg, incident.

factories and heavy engineering shops, thus generating employment, and there would be reluctance to write these off instead of generating projects with them.

THE SLINGS AND ARROWS OF NUCLEAR DEVELOPMENT

In fact, the most awkward aspect of nuclear power is not today presented by the power plants, brought back, one might say, from feverish growth to a more normal rhythm of growth, at any rate in comparison with the world's historical industrial evolution. The awkward aspect is the entire fuel cycle: from the mine to the storage of waste (perhaps excepting conversion to uranium hexafluoride for enrichment and the fabrication of fuel assemblies).

Uranium resources are not well known (see later discussion), and there is doubt whether they will suffice to sustain a long-term nuclear program without practically forcing us to have recourse to breeder reactors. If the earth's crust did contain sufficient uranium, could it be produced in time to supply an uncertain market, in which, moreover, a dreadful brand image results from quarrels between consumers (electricity companies or sellers of reactors) and producers accused of forming cartels.[5]

Enrichment seeks its path, unfortunately always an expensive one, and swings between penury and abundance in the long term. Reactors, too, in a certain sense are seeking their path: towards miniaturization and geographical dissemination, or towards continual gigantism leading to nuclear parks.

Downstream from reactors, difficulties and questions are growing, like a river widening as it reaches its mouth. Must irradiated fuel be reprocessed? And if so, where? Currently there is no commercial reprocessing plant in the world, with the exception of France's Cogema plant at La Hague. The result? Reactors in operation are accumulating irradiated fuel in their pools, and operators do not know what to do with it and are living under the threat (not all, but some of them) of being brought to a standstill because of lack of storage space for their used fuel.

The irritating problem of radioactive wastes still remains, a problem that for too long was underestimated. Evidently it was less glorious, less interesting, and less profitable than the construction of reactors. For radioactive wastes, bits and pieces of solutions exist, but not *the* solution, one validated by experiment, one that is acceptable and accepted. But such a solution now seems more and more likely to be a precondition for the acceptance of nuclear power by the public.

It is clear that those responsible for nuclear development have committed an error, which is now costing them dearly, in underestimating the importance (psychological probably even more than technical) of this very long-term handling

[5] In the Westinghouse affair, among others, the supplier sold, at low prices, uranium that it did not possess but hoped to be able to procure at a still lower price. However, the price of uranium in fact sky-rocketed, and Westinghouse refused to honor its contracts, which would have brought it to bankruptcy. This set off a whole series of legal proceedings, uncovering a number of more or less "accepted" practices.

of wastes. There is a supplementary paradox. The sense of pressure caused by fear of the imminence of a new oil crisis is not conducive to calmness and to acceptance of the delays entailed by research into this fundamental problem. It is indeed a fundamental problem, and it is rendered more difficult by the fact that it would take a thousand years or so to provide a conclusive demonstration that the problem had been solved.

Acceptance of the proposed solutions essentially assumes a certain degree of confidence both in probability calculus and in those putting forward these solutions.

It is curious to note that it is virtually only France that has put forward a coherent program for reactors and all stages of the associated fuel cycle. This relative success, however, is not without its problems because of the commercial exploitation involved. In effect, France has offered to process irradiated fuel for other nuclear countries embarrassed by America's lapse: Germany, Austria, Belgium, Finland, Japan, Sweden, Switzerland, and so on.

Such a quasimonopoly might appear lucrative, at least on paper, particularly as the customers are providing finance in advance, without interest, for the construction of the installations. The contracts are said to be very one-sided and the sums to be charged are practically left open.

But these are contracts of which it may be said, in a sense, that they do not have a "price" because they would keep open the door to numerous foreign nuclear programs threatened by a moratorium. Such an unusual operation, however, offers numerous grounds for concern. There are untried techniques to be perfected, risks to be taken (vigorously denounced by the trade unions in the La Hague plant[6]), and above all, the role of international "atomic trashcan" to be shouldered. Thus far, France has been the only country boldly to accept this role.

Cogema's customers can dispose of their storage problems of irradiated fuel and management of radioactive wastes for years, even for decades. In theory these wastes will ultimately be returned to the country of origin, but this return to sender probably will not be complete and will be limited to highly active wastes. The fate of the reprocessed uranium has not been decided at present. In many respects, therefore, this operation appears highly risky and premature, a vertible trap with future complications.

Apart from France, most other countries, like Japan, have made haste with much less speed. It is fair to say they chose to depend on the enormous potential of American industry for reprocessing. As it happened, this industry wandered and digressed until President Carter's bludgeon stroke of 7 April 1977 announcing a halt, *sine die,* to reprocessing projects (in reality already practically at a standstill) and to the development of sodium-cooled fast breeder reactors in the United States.

[6] The author has personally worked in a "hot laboratory," dealing with fission waste, among other things. Even at the level of a gram, or a curie of radioactivity, this is not an easy task. At the level of a ton, or a billion curies, with the "routine" of an industrial plant, one cannot help feeling a little bit scared!

THE IMPORTANCE OF REPROCESSING

Certain proponents of nuclear power opposed to Carter's policies have categorically declared that the reprocessing of irradiated fuel is indispensable. But a word of caution: the importance of reprocessing, as currently planned, should not be exaggerated or dramatized.

If the uranium-235 and plutonium recovered, possibly at a cost higher than their commercial value, are recycled in the same reactors (essentially light-water reactors, at present) they will permit a saving of about 20 percent of natural uranium and a corresponding proportion in enrichment services. This is not negligible, but:

- the percentage could be reduced by varying certain parameters of the cycle upstream (such as a tails assay in the enrichment plant)
- this 20 percent is not really significant in relation to world resources or reserves of uranium in the light of current knowledge (with a degree of imprecision of 1000 percent or more!)

Few countries have a Pu recycling program (Belgium, for example, is one of the few). Such countries, and others without a short-term program for plutonium fast breeder reactors, not only do not need plutonium but would not even know what to do with it. They would have to store it, with all the attendant difficulties, That is, unless it is for military purposes or, at the least, to keep options open.

However, reprocessing is indispensable in the development of a plutonium fast breeder reactor program. In this case reprocessing is vital, an obligatory stage. It is evident that the conventional defence of reprocessing offered by various countries will be proportional to their active engagement in a program of plutonium fast breeder reactors (backed also by commercial considerations).

Indeed, the Carter nuclear policy is aimed at putting into question the entire nuclear development of the Western world. The basis for this drastic change is the fear of proliferation of nuclear weapons all over the world through development of nuclear power programs.

FEAR AND NUCLEAR PROLIFERATION

We do not have to go back to the Flood but just to a few months after the apocalypse of Hiroshima, when the leaders of a victorious America, the nuclear superpower, were already asking themselves about the future of the power they had so tragically unleashed.

During the war, the three Anglo-Saxon allies (the United States, Great Britain, and Canada) had maintained a policy of strict secrecy (sometimes unilateral, by the way, to the benefit of the United States alone) and at the Quebec summit conference in August 1943 had concluded the first nonproliferation agreement concerning nuclear weapons. They agreed not to communicate any important information to a third party except by mutual consent.

In this policy of secrecy, they had in mind particularly the construction of weapons, including two areas they considered ''sensitive'': the enrichment of uranium and the reprocessing of irradiated fuel.[7] This defence against proliferation was made practically absolute by a policy of buying up all the uranium available in the Western world.

In March 1946, the Acheson-Lilienthal Committee solemnly warned the world that, if some international authority did not quickly take control of all national nuclear programs, civil or military, it would not be possible to prevent the proliferation of nuclear weapons. The phrases used are so topical that we must quote the most important ones:

> The development of atomic energy for peaceful purposes and the development of atomic energy for bombs are in much of their course interchangeable and interdependent. From this it follows that although nations may agree not to use in bombs the atomic energy developed within their borders, the only assurance that a conversion to destructive purposes would not be made would be the pledged word and the good faith of the nation itself. This fact puts an enormous pressure upon national good faith. Indeed it creates suspicion on the part of other nations that their neighbors' pledged word will not be kept. This danger is accentuated by the unusual characteristic of atomic bombs, namely their devastating effect as a surprise weapon, that is, a weapon secretly developed and used without warning. Fear of such surprise violation of pledged word will surely break down any confidence in the pledged word of rival countries developing atomic energy if the treaty obligations and good faith of the nations are the only assurances upon which to rely.
>
> Such considerations have led to a preoccupation with systems of inspection by an international agency . . . We have concluded unanimously that there is no prospect of security against atomic warfare in a system of international agreements to outlaw such weapons controlled *only* by a system which relies on inspection and similar police-like methods. The reasons supporting this conclusion are *not merely technical,* but primarily the inseparable political, social, and organizational problems involved in enforcing agreements between nations each free to develop atomic energy but only pledged not to use it for bombs. National rivalries in the development of atomic energy readily

[7] It is during reprocessing that strategic materials can be diverted from their normal use.

convertible to destructive purposes are the heart of the difficulty. So long as intrinsically dangerous activities may be carried on by nations, rivalries are inevitable and fears are engendered that place so great a pressure upon a system of international enforcement by police methods that no degree of ingenuity or technical competence could possibly hope to cope with them . . . We are convinced that if the production of fissionable materials by national governments (or by private organizations under their control) is permitted, systems of inspection cannot by themselves be made 'effective safeguards . . . to protect complying states against the hazards of violations and evasions.' If nations may engage in this dangerous field, and only national good faith and international policing stand in the way, *the very existence of the prohibition* against the use of . . . piles to produce fissionable material suitable for bombs would tend to stimulate and encourage surreptitious evasions. The effort that individual states are bound to make to increase their industrial capability and build a reserve for military potentialities will inevitably undermine any system of safeguards which permits these fundamental causes of rivalry to exist. In short, any system based on outlawing the purely military development of atomic energy and relying solely on inspection for enforcement would at the outset be surrounded by conditions which would destroy the system.

At the United Nations, three months later, Bernard Baruch presented a proposition aimed at putting all strategic materials and activities under international control, with severe penalties for violations, and escaping the veto of the Security Council. The execution of this visionary plan would have entailed a loss of sovereignty unthinkable at the time (including that of the USSR, which had been kept out of the Anglo-Saxon atomic effort during and after the war). It would still be unthinkable today. The Lilienthal-Baruch plan was rejected as the cold war gained impetus and, as Bertrand Goldschmidt put it, "with it vanished humanity's last chance to live in a world without the bomb."[8]

Instead of adopting an international authority possessing power over future nuclear industry and the responsibility for exploiting it and developing it in the name of all countries, the United States unilaterally adopted the MacMahon Act (1946). This act was an expression of the prolongation of bellicosity and isolationist secrecy. It served practically to paralyse all trade and international collaboration in the nuclear field during the ten years following World War II.

The reaction of other countries was to go ahead with their own nuclear development, both military and for peaceful purposes. The USSR and Great Britain, among others, later followed by France and China, broke the monopoly and thus upset the geopolitical balance (or imbalance!). In this respect it might be said that the real secret of atomic energy was that it could be liberated.

Seeing that it could not prevent national programs, the United States then decided to try to police them and, in any case, to restrict them to peaceful purposes.

[8] Such a chance is like virginity: you can only lose it once.

Isolationism was followed by a policy of liberalization, with perhaps the same degree of excess. This was the program "Atoms for Peace" put forward by Eisenhower in 1953, in which the United States offered, with conditions, to supply research reactors, enriched uranium (of which it had a monopoly), and technical assistance in the development of the peaceful utilization of atomic energy. In exchange, the United States required the right of inspection and control, permitting them to assure themselves that services and equipment were not being diverted to military ends. It is this control that enables the United States to forbid the processing of their enriched uranium outside the United States.

This policy culminated in the first Atomic Conference in Geneva in 1955 and in the signature of more than 30 cooperation agreements with numerous countries. It might be added, however, that at the first Geneva Conference and at those that followed it (and, in general, as regards the dissemination of scientific knowledge on nuclear subjects) the Americans have generally kept their lips tightly sealed about the enrichment of uranium and reprocessing.

France was the first to create a scandal, at Geneva in 1955, by publicizing its results in reprocessing. On the other hand, up to today, France has been much more discreet concerning the techniques of enrichment, whether by gaseous diffusion or chemical exchange.

It was partly to create an international body better adapted to supervise the guarantees of technological security and control of SNM that the International Atomic Energy Agency (IAEA) in Vienna was created in 1957. Today the Carter administration heaves a sigh, regretting the amount of information thus disseminated.

Profiting from the atmosphere of temporary detente between East and West, an additional step was accomplished with the signing of the Treaty of Moscow forbidding all nuclear explosions except those underground.

Next, in 1968, agreement was reached on the Non-Proliferation Treaty, ratified in 1970, by which the five nuclear powers[9] agreed not to help any other country manufacture any explosive device. Other signatory powers unilaterally and voluntarily renounced such manufacture and agreed that they would submit all their nuclear activity to checks by the IAEA inspectors.

Article IV of the treaty, which liberalises international exchanges, was one of the essential conditions for the treaty's acceptance by many countries (see box). Of course, this article is very important for developing countries without nuclear weapons; it was the price of their renunciation.

The Non-Proliferation Treaty (NPT) was said to have been abrogated or at least badly distorted by the Carter policy statement of April 1977 (to be discussed). Sigvard Eklund (Director General of the IAEA) was among the first to point this out. He and others did so at the May 1977 Salzburg Conference on Nuclear Power and its fuel cycle. Slightly earlier, this same point had been made at the Shiraz Conference in Iran on the transfer of nuclear technology, which opened practically

[9] The United States, the USSR, the United Kingdom, France, and China.

THE FAMOUS ARTICLE IV OF THE NON-PROLIFERATION TREATY

It was this article that "bought" the consent of the developing (nuclear) countries and that they consider jeopardized, to put it mildly, by the Club of London and President Carter's New Energy Policy:

1. Nothing in this Treaty shall be interpreted as affecting the inalienable right of all the Parties to the Treaty to develop research, production and use of nuclear energy for peaceful purposes without discrimination and in conformity with Articles I and II of this Treaty.*

2. All the Parties to the Treaty undertake to facilitate, and have the right to participate in, the fullest possible exchange of equipment, materials and scientific and technological information for the peaceful uses of nuclear energy. Parties to the Treaty in a position to do so shall also co-operate in contributing alone or together with other States or international organizations to the further development of the applications of nuclear energy for peaceful purposes, especially in the territories of non-nuclear-weapon States Party to the Treaty, with due consideration for the needs of the developing areas of the world.

at the same time as Carter made his declaration of 7 April 1977.

In fact, acceptance of the NPT has been less general than acceptance of the Treaty of Moscow, and some score of countries (including France, which nevertheless applied its principles) have refused to adhere to it or have not yet ratified it.[10]

Meanwhile there has been the "peaceful" Indian underground explosion in May 1974 — a political bombshell in the nuclear world — and contracts have been signed by advanced nuclear countries involving "sensitive" or "proliferating" equipment:

1. The contract between Germany and Brazil for eight reactors and all the installations for a fuel cycle complete with enrichment and reprocessing (certain experts have dared to say that Germany would thus make the bomb through the agency of Brazil.)
2. French contracts with Korea (cancelled)
3. French contracts with Pakistan (in the process of being cancelled as of mid-1978) for reprocessing factories.

Without doubt, the bomb that India exploded in the Rajasthan desert initiated a chain reaction among candidates for nuclear weapons: Brazil, Argentina (because

* Articles obliging countries with nuclear weapons not to supply these (or assist in their supply) to countries not possessing them and obliging these latter countries not to receive or manufacture such weapons or to seek aid in acquiring them.

[10] It might also be noted that Nixon and Kissinger offered to supply reactors to Egypt and Israel, although neither of these countries was a signatory of the NPT.

of Brazil), Pakistan (because of India), South Korea, and so on. It is this chain reaction and the Indian explosion that have probably precipitated "revisions" of the U.S. administration's previous positions. These are already effective, because the United States decided unilaterally to modify supply contracts, which greatly contributed to a regrettable deterioration of the international nuclear climate.

In the meantime, international commercial nuclear competition has become particularly active. Set off between 1955 and 1958 by the first sales of nuclear power stations by the British (Latina in Italy and Tokai-Mura in Japan), it was vigorously relaunched by the all-powerful American industry, strong from the success of the light-water reactors of its naval program and civilian program and backed by the policy of liberalization of Atoms for Peace.

Then came the threat of a shift from the reactor market to the market for installations. This was seen in the German-Brazilian contract and the Franco-Pakistani and Franco-Korean contracts. It was clear that international competition was increasingly putting at risk the objectives of the NPT. For this reason, the principal suppliers of nuclear material and equipment from East and West (first eight, subsequently fourteen suppliers) have formed the "Club of London," seeking to impose strict rules, that is, an exporting code that would strengthen rather than weaken the NPT.

More and more, the world is being divided up far from the ideas of Baruch-Lilienthal and the dreams — idealistic perhaps, but generous — of the first nuclear experts who saw in the atom the promise of cheap and abundant energy for everybody. There are two groups of players in this game. On the one hand, one finds the developed countries that have access to advanced nuclear technology (not necessarily military, as is seen in the case of Belgium or Sweden, for example). These recognize the well-founded fears of proliferation but hope to gain in international markets what their industry has lost at home through insufficiency of domestic markets and the slowing down of national programs. Such is the case of Germany, among others, where the KWU (Kraftwerk Union, a Siemens subsidiary) feels its very existence threatened by the slowing down of the German program and for whom the Brazilian program was like a whiff of oxygen.

It has not been easy, however, to find a compromise between international security and commercial requirements, where the smallest transaction is in billions of dollars. France, whose activity in international markets is sometimes questionable, has put forward the idea that perhaps candidates for "sensitive" installations (that is, for enrichment and reprocessing) might be prepared to renounce them if they were offered the possibility of turning to a certain number of serious suppliers (with the understanding that these suppliers would not renegotiate their contracts, as did the United States) who would guarantee free supply of fuels, enrichment services, and reprocessing. (France is a seller of these two services.)

On the other hand, there are the less developed countries (LDCs), underdeveloped both in nuclear and in other terms. They rightly feel themselves more and more manipulated by the developed countries and by what they call the "suppliers' cartel." Are they not being told — in a tragically false way and more and more for simply commercial reasons — that nuclear power is the key to their energy requirements, their well-being, their development, while in fact this way is poisoned both by its cost and by its difficulties?

One needs only to look at the slowness of the Indian program, though this is one of the most technologically advanced countries, to see how different the nuclear path is for an LDC. The IAEA in Vienna has contributed not a little towards the unwise belief in nuclear power in the LDCs by publishing unduly optimistic reports. Nuclear power can be of interest as yet to only a small number of Third World countries. But how is it offered to them today? With irritating practices accompanied by neocolonialism, which naturally increases their desire to become members of the nuclear club, too.

It is in this climate of economic and industrial uncertainty and of political confusion that President Carter dropped his "bomb" (which, however, was preceded by a number of warning signs). By unilaterally renouncing the program of development and commercialization of fast reactors and reprocessing of spent fuel, both *in their current form* (a point which is often forgotten), Carter was hoping to show the world the moral power of the United States. He also invited other nations to follow this example.

It was a naive step that Carter took, given the size of the interests involved. To date it has done nothing other than further to increase — as if there were any such need — international confusion and to add to the North-South impasse a profound misunderstanding in the camp of the industrialized countries, where hypocrisy flourishes.

This leads us to put the question: is there such a thing as "good" nuclear power, nonproliferating? It was to try to answer this question that the United States launched the International Nuclear Fuel Cycle Evaluation program (INFCE) planned to last two years (1977-1979). Most European countries agreed to participate in INFCE, but without enthusiasm and with the condition that no national program of the participating countries would be suspended before the program had produced its conclusions.

The object is to research a new type of reactor (possibly) and, above all, the associated fuel cycle, in which it would be impossible at any stage to recover and divert from their normal use any strategic materials for use in nuclear weapons. As this cannot be made absolutely impossible, systems are being sought that will make these diversions as difficult as possible.

The INFCE project is a think-tank such as the nuclear community has not known for over twenty years. But the atmosphere is no longer the same. Old projects are brought out of the drawer, such as the reactor with the varying neutron spectrum or the ever-attractive reactor using molten salts, and new propositions are being dissected, such as the use of uranium-233 doped with uranium-238 (the two isotopes needing expensive and complex installations for separation) or co-reprocessing, in which uranium and plutonium would be purified without ever being separated from each other. But the enthusiasm of the early years no longer exists.

Indeed, the fact cannot be hidden that as nuclear energy has grown older, it has got terribly heavier, has ankylosed. It is now known that the smallest achievement costs billions of dollars or francs and years of extensive and expensive effort. Everybody still remembers the abortive attempt to break through with commercial high-temperature reactors in 1975, despite the strength of its sponsors, Shell and Gulf, and despite the undeniable and recognized interest in the idea.

Looking further, these same big problems of the thermal reactors also confront fast breeder reactors. Will the "ideal" reactor even be discovered? Who will dare to develop it and put it into operation? Who will take the risk of operating it on a commercial scale? There is also the uranium-233 cycle, undeniably attractive but existing only on paper. There is no uranium-233. It would have to be manufactured in special reactors (as plutonium is made) and reprocessing and fabrication of fuel would have to be organized on a large scale. In short, this would involve a twenty-year program costing several billion dollars. Maybe the game is indeed worth the candle, but nuclear power has, in a way, become a victim of its own sophistication and gigantism.

All the foregoing considerations appear to be in vain, because there are plenty of other ways to acquire strategic material than through the cycle associated with power reactors. One can obtain plutonium with a simple research reactor — the way followed by India[11] — and relatively insignificant reprocessing equipment.

One can also obtain enriched uranium by centrifuging or by lasers, sectors about which the Carter program is curiously silent.

It might also be noted that Carter's promise to make available large quantities of enriched uranium for a "reinforced" program of light-water reactors has not convinced anybody. The American factories that would supply this uranium still have to be built, and potential customers distrust such an undependable supplier more and more. So plans for enriching plants flourish all over the world. There are projects in France, the United Kingdom, the Netherlands — and Brazil? — that could be augmented by Italian, Japanese, Australian, and South African projects, thus producing a result exactly opposite to the one aimed at.

Finally, the most that one can expect of any policy of nonproliferation, thirty years after the rejection of the Baruch-Lilienthal proposition, is that it will make the obtaining of strategic materials more difficult, and slow down proliferation. One may ask oneself whether this is any consolation, whether a world with twenty nuclear powers is better than a world with a hundred, and hope that any nuclear wars that may arise will remain localized.

Nuclear development could perhaps be limited in a general way to what currently exists. That is to say, with a possible and nonexclusive variant fuel cycle, it might be limited to light-water reactors, as President Carter has implied; but this

[11] With a research reactor supplied by Canada in the 1950s.

in turn brings up the acute question of uranium resources and, depending on the answer to this question, the problem of breeder reactors.

URANIUM RESOURCES

Initially, uranium was a very special product that had essentially only one buyer, the American government, and only a single, noncommercial use: the construction of nuclear weapons. With the birth and development of programs of electronuclearization, times have changed, and it is necessary to know how much uranium is or would be available. But as is the case with oil, alas, not very much is known about uranium resources.

If one believes the gross figures, there would be just enough uranium to last, at best, to the end of the century. This would mean that the supply prospects in the long term for uranium would be less favorable than those for oil, whose approaching end is being trumpeted.

In fact, however, uranium and oil are not strictly comparable: the search for uranium is just beginning, while oil exploration is over a hundred years old. Numerous experts and geologists are convinced that there is a lot of uranium remaining to be discovered and, to use a hopeful expression applied to coal, that there is a significant "potential behind the potential." Some people — such as the Ford Foundation, whose conclusions have strongly influenced President Carter's nuclear policy — even maintain that the world's potential uranium resources are so great that the development of breeder reactors designed to economize on uranium is no longer urgent.

Not everybody thinks uranium is so abundant. Such eminent geologists as Claude Guillemin, director of the Bureau de Recherches Géologiques et Minières, are relatively pessimistic. They cite the scarcity of rich deposits, the relative scarcity of medium or small deposits permitting surface working, and the possible abundance of large deposits but with low grade ore, difficult or impossible to exploit because of their depth, except at astronomical prices.

To get a better knowledge of reserves (both necessary and difficult), of resources, and of the "potential behind the potential," increasing efforts are being made by many countries. In Canada and the United States, programs are being developed for the evaluation of national uranium potential, major programs that will extend over several years.

The NEA (OECD) and the IAEA regularly publish data available on assured reserves of uranium and possible additional resources.[12] The latest study, published in December 1977, is interesting because since the previous estimates in 1975 the world price of uranium has increased considerably. From $6-15 in 1975,

[12] This is a category which in the McKelvey diagram (see the chapter on oil) will be counted as probable reserves. The NEA (OECD)-IAEA take into account two types of resources: reasonably assured resources (equivalent to "proved reserves") and additional estimated resources. Taking two price categories for each of these gives four categories altogether.

the cost of a pound of uranium oxide, U_3O_8, has risen to $40 and even more for certain spot contracts. Does this result from the existence of an uranium cartel? Or is it a slightly swollen effect of cost inflation? This increase is all the more significant, since the numerous programs of exploration stimulated by the 1973-74 oil crisis are now beginning to bear fruit. If the figures can be believed, the results of this exploration have nevertheless been fairly modest, but perhaps we should wait a year or two before making a final judgement.

In order to cater to inflation the NEA (OECD)-IAEA working group has revised its cost brackets upwards (for costs, not prices), with the lower range now being up to $30 per pound of U_3O_8, as compared with $15 per pound in the previous 1975 survey. This represents a jump of 100 percent. The next range is from $30-50 per pound of U_3O_8, compared with $15-30 two years previously.

The estimates call for some comment:

- Proved reserves, or, to use the jargon of the report, reasonably assured resources considered economic, have risen (for the Western world, excluding countries with centrally planned economies) from 866,000 tons of uranium in 1973 to 1,080,000 tons in 1975 and 1,650,000 tons in 1977 (these figures being deduced from production of about 20,000 tons per year). The figure has practically doubled over four years, which is quite remarkable.
- The quantities in the higher price categories are generally less than in the lower price categories, while the economic theory of natural resources would suggest the opposite should be the case. However, this phenomenon simply reflects the fact that there has been less interest in estimating these resources, as is only to be expected. The figures for these categories are thus of limited significance.
- The additional estimated resources in the two price categories together amount to 2.1 million tons of uranium, as compared with 1.68 million tons two years previously. The experts think this progression is a modest one. The principal

TABLE 2-1.
The Principal Uranium Countries

	RESERVES		TOTAL RESOURCES	
	× 1000 t U	% world	× 1000 t U	% world
United States	523	31.70	1,696	39.53
Canada	167	10.12	838	19.53
South Africa	306	18.55	420	9.79
Australia	289	17.52	345	8.04
TOTAL	1,285	77.89	3,300	76.90
France	37	2.24	95.9	2.24
France and African associates	225	13.64	354.9	8.27
WORLD TOTAL (all countries)	1,650		4,290	

increases were found in the United States and Canada, which are counting on finding still more additional resources in known uraniferous districts.

- A major part of the world's uranium is found in a relatively small number of countries. Excluding the countries with centrally planned economies, these are the United States, Canada, South Africa, Australia, and, if one includes its African associates (Central African Empire, Niger, and Gabon), France (see Table 2-1). The first four countries named alone control or possess 78 percent of the economic reserves of reasonably assured resources (under $30 per pound of U_3O_8) and about 77 percent of total resources (that is, the four categories taken together).

These data admittedly cover only a part of the world's uranium resources. In order to go beyond this, the NEA (OECD)-IAEA have started a joint program of international collaboration with a view to determining the world's uranium potential, this being entitled IUREP (International Uranium Resources Evaluation Program). After eighteen months of work involving a score of experts, IUREP has presented its first results, summarized in Table 2-2.

If one takes the higher value for the world estimate and adds the reserves and resources from Table 2-1 plus cumulative production up to the end of 1979, one arrives at a total of about 26 million tons of uranium. Curiously enough, two completely independent additional estimates, using quite different methods — Perry and Belototsky using a geological analogy and Skinner looking at uranium's abundance in the earth's crust — also arrive at this same total of 26 million tons. Not too much importance should be attached to these coincidences, but a figure of

TABLE 2-2.
Speculative Resources Listed by Continent

Continent	Number of countries	Speculative Resources (million tons U)
Africa	51	1.2- 4.0
America, North	3	2.1- 3.6
America, South and Central	41	0.7- 1.9
Asia and Far East*	41	0.2- 1.0
Australia and Oceania	18	2.0- 3.0
Western Europe	22	0.3- 1.3
Subtotal	176	6.5-14.8
Eastern Europe, USSR, People's Republic of China	9	3.3-7.3**
Total	185	9.8-22.1

*Excluding the People's Republic of China and the eastern part of the USSR.
**The potential shown here is "Estimated Total Potential plus production." As data are not available to the Steering Group to indicate the size of presently known resources, Speculative Resources cannot be assessed separately.

about 26 million tons of uranium may perhaps be taken when considering the long-term perspective, particularly in view of our current limited knowledge.

NONCONVENTIONAL URANIUM RESOURCES

It is interesting to examine, as we will do for oil in a later chapter, the potential of nonconventional uranium, including the "inexhaustible" sources sometimes mentioned: granites, sea water, and so on. However, these possibilities might materialize only in the course of the next century, or maybe never.

To date targets for prospecting have generally been deposits with a yield of 0.1 percent of uranium or more[13] (1000 parts per million, ppm). This figure is curiously similar from country to country despite the different types of mineralization (such as sandstone in the United States and Hercynian veins in France). The consistency of yields, incidentally, reinforces the thesis of proponents of a big potential that there are plenty of finds still to come.

There is little knowledge of deposits with lower ore content, although some of these are beginning to claim some attention because they can be regarded as extensions of more classical deposits, relatively well known. Additional low content deposits may be found in new contexts, nonclassical and not yet prospected. A number of these deposits, with an ore content of 0.1 percent to 0.05 percent (1000 to 500 ppm) could be exploited, given a price range of $40-60 per pound U_3O_8 (1978 dollars).

In addition to these possibilities must be added that of uranium as a by-product from the production of phosphoric acid by the wet method and as a by-product from lixiviation[14] of copper ores. With world production capacity exceeding 20 million tons per year, phosphoric acid represents a maximum theoretical potential of 6400 to 8400 tons of uranium, a third of current world production. A commercial by-product installation recently entered service in Florida, and numerous other projects are under way since uranium prices have risen so sharply. (Projects exist in the United States, Canada, Spain, Morocco and Israel, among others.)

Carrying on down the scale of grades, it may be noted that two nonclassical techniques for in-situ recovery from certain low-grade ores are being studied: lixiviation in situ and bacterial lixiviation. At the bottom of the scale are the very low grade ores, the quantities of which in absolute, and theoretical, terms might satisfy man's energy needs for thousands of years. But it is not possible to say today if they will ever be exploited, that is to say, if they will ever be technologically and economically recoverable.

We may cite:

[13] The enormous Rössing deposit in Namibia (150,000 tons of uranium) is an exception, with a content around 0.04 percent) (400 ppm).

[14] A method of treating ore analogous to leaching.

- black marine shales. Uranium content in these usually varies between 30 and 60 ppm and can occasionally reach or exceed 300 ppm, as in the central part of southern Sweden, where there is a pilot plant at Ranstad. Incidentally, the Swedish government has recently refused to enlarge this pilot plant, for ecological reasons, and has asked the promoters of the project to consider the recovery of all the minerals contained, not just uranium. Other shales, such as those in the Chattanooga region of the Unites States, also contain uranium (50-60 ppm). But to exploit them would mean an enormous mining impact on the environment. For each ton of uranium it would be necessary to extract 20,000 tons of rock, not counting sterile rock and overburden.
- igneous rock. "Burn the rock" was an idea of Weinberg put forward a decade ago, the uranium in the rock being able in principle to feed breeder reactors "in perpetuity" (for ordinary water reactors it would be still better to return to coal and burn it up to the last recoverable atom!). Most igneous rock contains 2-4 ppm of uranium, sometimes more (in the case of certain granites and syenites). In reality there is as far as we know no program with such a diluted source in mind.
- sea water. Although this is still more dilute — three thousandths of a part per million, a thousand times less than in granite — uranium in sea water has continued to attract attention for a number of years, because a liquid is relatively much easier to manipulate than a solid and because the theoretical potential is simply gigantic: 4,000 million tons of uranium. Numerous research programs have been started, in the Federal Republic of Germany, the United Kingdom, Italy, and Japan, and also some in France. The technical feasibility of some processes has been demonstrated in the laboratory. But the problem of manipulating enormous quantities of sea water remains, and this has to be solved if sea water extraction is to become practical.

In conclusion it can be said of uranium, as of petroleum, that our knowledge of resources is largely insufficient to be able to serve as a basis for long-term energy policies. One way of appreciating this problem is to compare available reserves and anticipated resources with requirements for the various nuclear programs, even if the forecasts were continually revised downwards. In this respect it may be noted that despite successive reductions of programs the demand for uranium has actually increased, largely because the United States have revised their contractual policy on enrichment. From customers, they demand firm orders for years in advance. They have also revised their storage policy and the working conditions of their enrichment plants to minimize energy consumption. This has suddenly inflated the demand for natural uranium and induced a fear of scarcity. Electrical managers have protected themselves against possible scarcity by extending the length of their uranium supply contracts (sometimes for the entire planned lifetime of the power plant). Naturally, all of this has increased tension in the uranium market and has contributed to the sharp rise in prices.

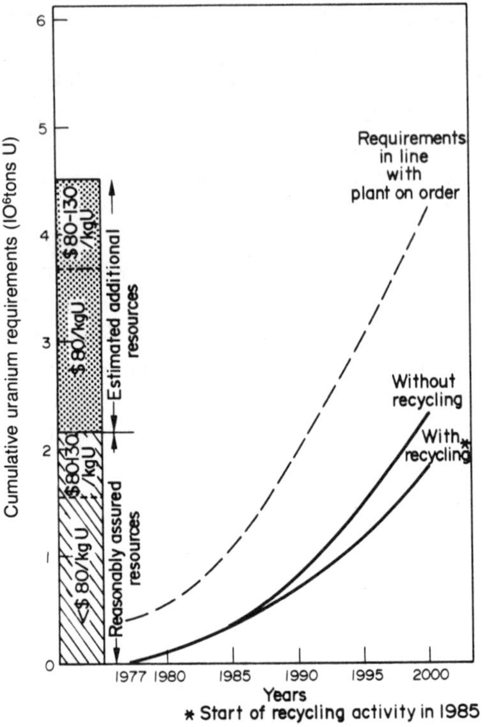

FIGURE 2-1

Factors of this type and uncertainty about long-term nuclear programs have led, as we have seen, to estimates of uranium needs varying by 100 percent. Hence, the forecasting task of the mining industry has become difficult, which increases the risk of overproduction or underproduction. Such a phenomenon of oscillation is familiar for numerous raw materials. Uranium was hoping to escape it.

Figure 2-1 shows the NEA (OECD) and IAEA forecasts for nuclear power growth in the light of the hypotheses discussed above. We show only the low hypotheses. A nuclear revival would push the curves up considerably and would increase the problems. In the figure, the first two curves (solid lines) show the growth of demand and estimate the normal supply of uranium. The case "without recycling" means either without reprocessing or with reprocessing but without immediate re-use of the plutonium or uranium. This case shows that reserves or reasonably assured resources at under $30 per pound and under $50 per pound (those known in fact or virtually assured today) are just about sufficient to last till the year 2000.

The broken line above shows requirements in the case of "panic" stocking, or simply prudence, when the electricity managers order or reserve in advance for each 1000 MW(e) plant the 5,000 tons of uranium it will need. In this case, known reserves would be committed between 1985 and 1990.

These curves, and the higher curves that would correspond to an accelerated world nuclear program, show the urgent need to intensify exploration programs. It is essential that both the estimated resources and the assured reserves be increased. One can have reasonable hopes that these increases could be achieved, provided that the necessary decisions are taken in time. If this is done, the supply situation could be adequate *from now until the end of the century*.

Things will begin to change after the year 2000. If nuclear power expansion continues, even if it follows only the low hypothesis, uranium requirements will become more and more important and towards 2030 could reach a cumulated level of 7 to 10 million tons. This approaches the limiting values estimated by certain experts for accessible uranium resources at a reasonable price, although possibly this is a slightly pessimistic view that underestimates technological potential and progress in methods of extraction and recovery.

In other words, within a horizon represented by half a century one is faced with the principal alternatives: either to accept the possibility of an appreciable jump in the cost of uranium (to a level of $100 per pound U_3O_8 or more, which would naturally have repercussions on the price of electricity), or to utilize uranium in a more efficient way, including in breeder reactors.

A progressive introduction of breeder reactors from 1985 onwards, as assumed in some studies, is undeniably optimistic; the world does not seem to be aiming toward such a rapid introduction at present. Nevertheless, two things should be remembered: the introduction of breeder reactors will lead to a slowing down in the growth of demand for uranium. The cumulative demand curve will flatten. But the later and the more slowly breeder reactors are introduced, the later will this flattening occur, and the higher will be the cumulative value of this demand for uranium.

This finally brings us to the detailed analysis of the problems of the breeder reactor.

BREEDER REACTORS

The nature of the breeder reactor is little short of miraculous: a machine that produces more fuel that it uses. Opponents of the breeder reactor will say that it is not a miracle but a curse, and that this machine is infernal. It is difficult to fathom why the breeder reactor has become the focus of public opposition to nuclear power, and plutonium the symbol of humanity's shame. Between water reactors and breeders there is only a difference of degree, not of nature. Let us try to see clearly. Without illusions! For these views I know that I shall be attacked by both opponents and proponents of the breeder reactor.[15]

Although there are thermal and fast breeders, and among fast breeders various

[15] It is faint consolation to know that for once they will be in agreement against something.

possible types (GCBR or Gas Cooled Breeder Reactor, LMFBR or Liquid Metal Fast Breeder Reactor) only the LMFBR has been developed so far to the prototype or precommercial stage. Most of our comments will thus apply to the LMFBR, referred to as the fast breeder (or simply, the breeder).

The first question one may ask is whether the breeder has such economic advantages that it is bound to impose itself on the market. It seems the answer is "no, it is not *bound* to." As stated earlier, it has become more and more difficult to introduce a new system, whatever it be; and in this case we are looking at a new type of reactor *or* the fuel cycle associated with it (such as reprocessing or refabrication).

This is a paradox working against the breeder reactor. One no longer believes in the figures on paper and in the table-thumping of breeder promoters seeking to prove the economic advantages of their system. However, truly credible figures could be obtained only if the system was effectively developed. This is the problem of the chicken and the egg, and the reason that, as time passes, there is an increasing tendency for nuclear power to stick to the existing system of the light-water reactor.

Moreover, in the case of the fast breeder reactor, all the calculations agree in showing that this type is appreciably dearer in capital cost than a light-water competitor of equal power: (only) 20 to 35 percent more, say some (the French, among others); 50 percent more, say others; while wicked tongues say 100 percent more, or not far from it (at any rate for the first X units, with X varying from one to infinity!).

The proponents of fast breeder reactors promise that the cost of the associated fuel cycle will be appreciably less than that of light-water reactors.[16] In the absence of valid industrial experience, this remains to be demonstrated . True, the cycle is effectively simpler and involves fewer stages, but on the other hand, each stage is more complicated than those of the LWRs.

Of course, the experts do not agree on the value, but they are sure about one fact: uranium will have to be much more costly than it is today (twice as much, three times, five times?) before fast breeders compensate for their additional capital cost and become competitive on a purely economic basis.

The fast neutron breeder system currently proposed is more complex than the thermal neutron systems. The time has passed when what was new and complex was best. Complexity also affects safety. No doubt, fast breeders can be made as safe as water reactors, but this must involve a good deal of additional effort and entail major precautions; both of these add to plant cost. This is the price to be paid for high performance.

Finally, the necessity for reprocessing and the need to handle almost pure plutonium (with the present state of the system) add to the possibility of nuclear

[16] Remember that the price of the kW.h is made up of amortization of capital, costs of the fuel cycle, and running costs. The capital investment in the reactor is the most important item, at 60-70 percent of the whole.

weapons proliferation, even if this is certainly not the simplest way to obtain such a weapon. In the light of all these considerations, to be able to master and utilize the system is a certificate of ability to do anything nuclear.

If there are no economic reasons for introducing fast breeder reactors, are there any other reasons? The answer will vary according to whether one is thinking on a global scale — but can one do so in the absence of world management of resources? — or on a regional or national scale. In both cases, the answer naturally depends on uranium resources, and future demand.

The Global Scale

It suffices to turn again to the curves in figure 2-1 to see that the introduction of breeder reactors *seems* to be ultimately a necessity. But when is ultimately? How urgent is it? *These are the questions.*

Are uranium resources limited? Once more, one finds the same problem. One does not know, with certainty, what the resources are. Their scarcity would increase their cost, and the breeder reactor would become competitive. Even without this information there could and ought to be some ethical principle concerning the management of natural resources. The breeder reactor is sparing of them, and that is its big advantage.

A first reason for acceleration could emerge from the demonstration that uranium resources are not very large or are not very accessible. Access to reserves is becoming an increasing preoccupation of the mining industry. The mines, particularly open-cast ones, are destructive even if only temporarily, thanks to subsequent restoration of the land. Note that 95 percent of the world's ores are extracted in quarries, and there are increasing obstacles to this type of mine.

Let us not forget the definition of reserves: they must be economically *recoverable*. In the United States, where the federal "reserves" (in the sense of reserved land) are growing more and more rapidly, access to resources has become a major problem. Australia is another example. This trend will continue, and humanity perhaps will prefer to die poor on its treasure rather than to exploit it.

A second cause of acceleration (in our opinion potentially the most important) would be the resumption of nuclear programs. We are not judging the probability of resumption, but only the consequences. Resumption could lead to a galloping consumption of uranium, as did the programs way back in time. On analysis, the link becomes evident: if energy consumption mushrooms and if nuclear power supplies an important part of it, then the breeder reactor will become the favorite tool in major programs.

In considering possible reasons for braking the breeder programs, let us begin with the extreme case: its bad press may lead to the rejection of the fast breeder reactor. Today this possibility cannot be ruled out completely.

Contrary to the case for acceleration of breeder programs due to a relative scarcity of uranium resources, let us assume resources are found to be more abundant than currently estimated, for instance with the discovery of new Australias. This would reduce the urgency of the need for the breeder reactor. There could be additional resources of conventional uranium, or new methods of extraction (in situ lixiviation) or a technological breakthrough in recovery of uranium from sea water. Remember that the study by the Ford Foundation, which inspired President Carter's policies, is staking much on the possibility of additional resources.

Other braking factors to consider are:

- Light-water reactors have not exhausted their potential for improvement. Their fuel burnup,[17] among other factors, today about 30,000 MWd/t, could go up to 40,000 or 50,000 MWd/t, which would reduce the cost of the cycle (reducing the theoretical advantage of the breeder reactor) and also the cost of uranium requirements.
- The success of thermonuclear fusion, apart from the competition it would provoke in the energy-producing sector, could open the way to the direct production of plutonium from natural or depleted uranium or the production of uranium-233 from thorium.[18] These fuels would permit the use of light-water reactors or any other burner reactors practically ad infinitum.
- To round off the breeder reactor picture, it is worth remembering that fast breeders, using sodium and plutonium, are not the only breeder reactors possible. There are others, some having the advantage of working with thermal neutrons (less dangerous) or with the cycle of thorium-uranium-233. Among others, one could cite (with a certain nostalgia, for I was warmly partisan to it) the reactor system using molten salts. Putting on my nuclear cap for a little while, I would say this type of reactor has many advantages over the fast breeder, including those of simplicity and safety. Hanging up my cap again in my souvenir cupboard, it is evident that everything said earlier about the difficulties of introducing a new system can apply, unfortunately, to the molten salt system as well, to the letter.

The Regional Scale

The notion of uranium scarcity takes on a new significance the moment one ceases to think in global terms. Certain countries, including European ones, are big consumers of energy but particularly poor in resources, whether fossil of fissionable. France, a leader in the world development of fast breeders, is quite a

[17] The fuel burnup corresponds to the quantity of energy one can produce per ton of fuel loaded into the reactor. It is measured in Megawatt days per ton or MWd/t. For example, natural uranium-gas-graphite reactors have a burnup of from 3000 to 5000 MWd/t; the present generation of water reactors have a rate of about 30,000 MWd/t; for fast breeder reactors, the expected rate is in the neighborhood of 100,000 MWd/t.

[18] See the following chapter on thermonuclear fusion.

good example (although its definitive inventory of coal, oil, and gas has yet to be made). For such countries, spurred on by the hope, or the carrot, of "energy independence," the breeder has everything to offer, or almost everything.

Here, too, it can be seen that the urgency depends on the rate of growth of future energy consumption, and the relative shares of nuclear power and electricity. In the case of France, taking about a hundred 1,000 MW(e) nuclear plants as coming into operation between now and the year 2000, this would involve using some 500,000 tons of uranium (5,000 tons per 1000 MW(e) for the lifetime of each plant). This represents five times all the national reserves and estimated additional resources (up to $130 per kg of uranium). This requirement is practically inevitable, anyway, for even if fast breeders become commercial in 1985, which is by no means certain, their effect would only make itself felt to any extent in the first decades of the next century. Until then there would simply be a transfer from dependence on oil imports to dependence on uranium imports, admittedly not from the same countries.

Two further points may be made about the French situation (the other European countries being effectively a few years behind, despite their verbal protestations):

- It is clear that the motive force behind fast breeders is the French Atomic Energy Commission, the research organization that developed them. The position or the interest of Electricité de France is less clear than its official statements might suggest. It must be remembered that one of the reasons put forward by EDF for its conversion to the light-water reactor was the practically worldwide adoption of this type, thanks to its relative reliability and the large amount of international knowledge concerning it. It may well be asked what will become of this argument following the adoption of the first fast breeders and, *a fortiori,* of electronuclear units that look like being the biggest in the world.
- Phénix and the Superphénix project are undeniably remarkable national products, and the legitimate pride of their creators and promoters is understandable. A commercial breakthrough in foreign markets would represent a culmination of their success. But it appears that these foreign markets, whether industrialized countries or developing countries, are not ready for these products. It is to be hoped for France's sake that the current forcing of these reactors will not result in a Concorde-type situation — the comparison is sometimes made — but rather to a Vandellos-type situation (a gas-graphite plant sold to Spain at very advantageous conditions, shortly before the type was finally abandoned). The surest course, and perhaps the hardest, will be to make haste slowly.

CONCLUSIONS

Having arrived at this point, two questions seem important in summing up: is nuclear power going to come to a halt? Is it necessary to halt nuclear power?

As for the first question, there are certainly several factors tending in this direction:

- United States policy. The significance of the United States in world nuclear development must not be underestimated. In the nuclear sector, United States capacity represents 50 percent of installed world capacity, and the United States perhaps has 80 percent of the scientific and technological knowledge in the nuclear field (if it is measurable). President Carter's policy in the nuclear power domain — as in many others besides! — is ambiguous. For the moment, his negative decisions (on reprocessing, on breeder reactors, and on the export of technology, equipment, or strategic materials) have outstripped by far his positive intentions, which have remained purely verbal (including the ''re-launch'' of water reactors). In the United States, the nuclear industry feels upset. Accustomed to receiving forty orders per year, it is now getting only three or four, and in recent years cancellations have often exceeded new orders. For how long can industry withstand the blows? If it surrendered, subsequent recovery would be particularly difficult, if not impossible.
- In Europe the situation is not much brighter, except in France, where Framatome seems to be an island of nuclear prosperity.
- Public opposition is not diminishing. However, it is certainly not hardening as much as convinced opponents of nuclear power would wish: one may conclude this by looking at referenda in six American states or recent opinion polls in Sweden and elsewhere. As regards nuclear sites, a crucial problem is arising, the guerilla.

Taking the broad view, there are some factors that lead one to believe in the survival of nuclear power:

- First of all, there is its intertia. As just stated, there is 100 GW(e) installed and another 200 GW(e) under construction; there is an industry heavy with investment; and there are customers, the electrical managers, remaining faithful — unhappy and sorely tried, but faithful.
- Its state of development. It is a ready technology, and its possible replacements will be ready only in the future, or only under certain conditions.
- Its performance, which is improving. It works, it produces, and the catastrophic accidents predicted for it have not become reality. Harrisburg was not a real catastrophe; the most serious aspect of the incident was the inaccurate statements made about it.

Between these opposing forces there is the unknown factor of energy demand: if it were extremely weak, nuclear power could be reduced to a marginal amount, even to the status of a bad memory. If demand were moderately weak, the nuclear power situation would be debatable. If energy demand runs strong, nuclear power would be strengthened by solid arguments in its favor, and there would be nothing to debate; not only nuclear power but also breeder reactors would be imposing features in the market place.

The oak or the reed? Will nuclear power survive the tempest? The foregoing list of opposing influences does not yet tell us the answer. It does bring us to the second question, however. If nuclear power is the oak, if it survives will it be necessary to take an axe and chop it down? In other words, is it necessary to halt nuclear power?

3

THE THERMONUCLEAR "LEAP"

With thermonuclear power, the race towards the terawatt could make a new leap forward. This is to say, while waiting for SOLFUS in some form or the domestication of antimatter — about which practically nothing is known! — thermonuclear power is presently the most scientifically oriented and the most sophisticated of the possible sources of energy — and the most complicated, so complicated that nobody has to date been able to demonstrate its viability, even on an experimental level.

The principle is marvellously simple: instead of splitting the nuclei of heavy atoms (the phenomenon of nuclear *fission*), one fuses nuclei of light atoms (the phenomenon of nuclear *fusion*), operations which in both cases release huge quantities of energy. Because of the repelling positive charges of the light nuclei, they have to be enormously accelerated for their collisions to be effective and to result in the formation of a heavier nucleus. The best way of attaining such enormous speeds is by heating, that is to say, thermal agitation (whence the term thermonuclear). A simple calculation shows that the necessary temperatures are of the order of several tens of millions or hundreds of millions of degrees. With these figures the simplicity stops, and the difficulties commence. Such temperatures exist only in the stars — a stimulating idea for scientists! — or at the heart of an atomic explosion, for which reason the hydrogen bomb is to date the only successful example of thermonuclear power in practice.

Remember that as temperature increases, bodies pass from a solid to a liquid state, then become gaseous, and finally turn into plasma, more or less ionized (the fourth state of matter), constituted of negative electrons and positive ions or atoms stripped of electrons. As the temperatures used or obtained on earth never exceed 5000°C (and are often well below this), one can get some idea of the technical challenge presented by fusion. This challenge, to put the sun in a bottle, seems so formidable that one may think, one *must* think that the advantages associated with such a form of energy would also have to be enormous to justify such an enterprise. In fact, if one forgets the stock image — which scientists, grown-up children that they are, do not detest! — of free and inexhaustible energy, the advantages of thermonuclear fusion compared with nuclear fission and breeding are appreciable and certainly justify reasoned and reasonable exploration. These advantages are essentially:

- very large fuel resources, almost inexhaustible in fact, if the deuterium-deuterium[1] reactions can be mastered. For the time being, a bold attack is being made on the simpler reactions of deuterium-tritium. The latter is derived from lithium, of which reserves are very roughly similar to those of uranium.
- greater safety. No "runaway" is possible, appreciably simplifying the problem of siting.
- the possibility of less radioactivity, perhaps 10 to 100 times less, with radioisotopes having a shorter half-life. Problems of cooling after a shutdown and longterm surveillance are also simplified.
- last but not least, no problems of proliferation or of strategic materials, *in principle*. But this is true only in principle, because the fusion laser could be a shortcut to the H-bomb. (Indeed, with nuclear power, one cannot get away from it.)

These advantages, which seemed enormous a few years ago, have been somewhat eroded by time, which has revealed huge difficulties in practice. It is in fact impossible to say today what the prospects of this system actually are. But it is desirable to pursue efforts further, encouraged by some undeniable progress.

But where is fusion today?

Apart from the most important parameter, temperature, which must be several hundred million degrees, two other parameters play a fundamental role in the physics of fusion: the density n of the particles reacting in the plasma and the time t during which the reaction can be maintained before the products disperse. The British physicist Lawson showed just over twenty years ago that in order to produce more energy than is used — the least one can ask of an energy machine! — the product nt must exceed 10^{14}. As fusion likes only extreme solutions, two courses appear possible:

- "slow" fusion, where t is of the order of a second and n consequently of the order of 10^{14} particles per cm^2, which corresponds to a very good vacuum. To achieve such confinement times, magnetic fields of complex shape must be used, around which the charged particles circulate like moths attracted by a light. This is the most explored route, with various machines with strange names like Doublet, Stellarator, Tokamak,[2] rivalling each other in the amount of

[1] Deuterium, or heavy hydrogen (D), exists in nature in the ratio of one atom per 6500 atoms of normal hydrogen. Tritium (T), or superheavy hydrogen, does not exist in nature; it is a beta-emitter, with a half-life of about 10 years.

[2] Initially there was even a Perhapsatron.

intelligence and knowledge invested in them;

- "rapid" fusion, where t is of the order of nanoseconds (1 second $\times$ 10^{-9}) and n of the order of 10^{23} particles per cm^2, corresponding to superdense supersolids that can only exist in a transitory state for a minute fraction of a second, needing gigantic instantaneous power of the order of a terawatt for their creation and ignition, power which only lasers not yet in existence may some day perhaps be able to provide.

SOME ASPECTS OF SLOW FUSION OR MAGNETIC CONTAINMENT

As we cannot go into all the problems of fusion, we shall restrict ourselves to a few main points, seeking to put fusion into perspective as compared with other new forms of energy, limiting ourselves initially to D-T reactions.

The reaction between the deuterium and tritium nuclei produces helium and a very fast neutron which contains 80 percent of the energy liberated. These are very energetic neutrons which, through reactions with the structure of the thermonuclear reactor, produce respectable quantities of radioactive material (indirect products of fusion, therefore, and not direct products as with fission). These very fast neutrons are slowed down in a lithium envelope that captures a part of their energy and also breeds tritium through transmutation (the lithium atom cannot digest the neutron it absorbs and turns into tritium and helium). Consequently, D-T fusion must face two "environmental" problems, which initially there was a tendency to minimize: the manipulation of large amounts of tritium and the radioactivity induced in the reactor's structure.

Under conditions of normal operation, the biggest threat seems to be associated with the danger of a tritium release. Although at any given instant the amount of tritium involved in the reaction is only a few grams (less than a milligram per m^3), on the other hand there will be between one and ten kg tritium in the lithium envelope, assuming an installed capacity of 1000 MW(e), corresponding to 10 to 100 million curies[3]. Tritium, like ordinary hydrogen, has the annoying property of being able easily to pass through hot metallic walls. The thermonuclear reactor will thus offer two ways for the tritium to diffuse through to the atmosphere in normal operation: through the circuit of the lithium envelope or through the cooling circuit for the production of energy with a steam cycle (by diffusion through the walls of the heat exchanger).

Currently there are no norms limiting releases of tritium to the atmosphere from thermonuclear reactors. However, if the United States norms for tritium releases from ordinary light-water nuclear reactors were applied, the degree of leaktightness of the lithium and heat exchanger circuits would have to be of the order of 99.999 percent. This seems technically *possible,* but not too easy to achieve. And what would it cost? Indeed, the norms for tritium should be defined taking both

[3] The curie corresponds to the radioactivity of a gram of radium, 3.7×10^{10} disintegrations per second.

fission and fusion reactors into account.

As regards the induced radioactivity, the levels might reach between 1 and 10 billion curies per 1000 MW(e), which is not negligible. In fact, what counts is not the level but the nature and lifetime of the induced radioactivity. This essentially depends on the structural materials chosen as containment for the lithium envelope: if these materials are made of vanadium, for example, the periods of "surveillance" would be relatively short, about ten years; for nickel steels the periods would be in the neighborhood of fifty years; for niobium alloys (interesting for their good characteristics at high temperatures, among other qualities) there would be hardly any difference from the products of nuclear fission. But the scientists are relatively confident — the economists have yet to share their optimism! — about being able to avoid the constraints of long-term or very long-term waste management, currently a problem weighing heavily on nuclear fission.

There is another important point not always made explicit in relation to controlled thermonuclear fusion: "thermal pollution." As stated, about 80 percent of the energy produced is carried away by the fast neutrons produced. Because these are not electrically charged particles, there is for the time being no alternative to recovering their energy in the form of heat, which brings us back to the usual problems. What about the other 20 percent carried by the product nucleus with a positive charge? After slowing down in an electric field, its charge could be captured on an electrode (using the opposite working principle to that of the particle accelerator). Because this is only 20 percent of the liberated energy, the impact of such conversion could be only marginal, and the bulk of the energy liberated will be recovered by having recourse to the classical thermodynamic cycles.

If one adds to this the energy requirements needed to initiate the reaction, the magnetic containment, and so on, one comes back — and this is rather vexing, after having gone to so much trouble to manipulate such high temperatures — to a final efficiency of the order of 30-40 percent, as with nuclear fission or "common" coal. And of course one cannot avoid the remaining 60-70 percent finally reaching the atmosphere!

Research has shown both the advantages and the disadvantages of D-T reactions. One may of course ask if the fusion of other light elements (such as deuterium, helium, lithium, beryllium, or boron) might not free us from some of the problems linked with tritium or the necessary transition to the production of "conventional" heat, possibly by aiming at the production of charged particles whose energy could be directly recovered in the form of electricity. Also, new cycles — including the D-D cycle — might possibly liberate us from the limitations of lithium supply (on a scale of thousands of years, it is true). But to bring about such cycles temperatures will be needed that are closer to a billion degrees than a hundred million degrees (a minimal difference on paper, perhaps, but *not* in the laboratory) and the *nt* product, density of particles (*n*) times containment time (t), will need to be about five times higher than those still being dreamt about today. For the time being, these projects can be consigned to the realm of wishful scientific thinking.

Is is possible today to get an idea of the cost of thermonuclear power, or its possible future in competition with other sources of energy? The answer is: not a definite one. The tendency to increased capitalization is an interesting pointer, which shows a certain parallel between thermonuclear power and the solar power that it is seeking artificially to imitate. The swing in the ratio of capital cost to fuel cost, already strongly evidenced in the move from fossil fuels to fissile fuels, would be accelerated even faster with thermonuclear fusion. Investment costs might be of the order of 90 percent (as against 70 percent for nuclear plants, and some 30 percent for the traditional fossil fuel plants), running costs about 10 percent and fuel costs 1 percent. It is only a step from 1 percent for thermonuclear fuel to 0 percent (in theory) for solar fuel. Indeed, the similarity does not stop there: solar and thermonuclear (slow fusion) power both use low-power densities (of the order of 7 kW per "litre of plasma" for fusion, as against 500-1000 kW per litre for a plutonium breeder reactor), which leads to relatively high specific investment in materials.

The last point to look into is the possible timetable for controlled thermonuclear fusion using magnetic containment. Currently the best-performing devices are reaching — or even exceeding — one of the critical parameters (temperature, density of nuclei reacting, or time of containment), but not all three simultaneously. The Princeton PLT and the Soviet T-10 (Tokamak 10) have reached a containment time of a tenth of a second; the Oak Ridge ORMAK (USA) and the French TFR at Fontenay-aux-Roses have exceeded 10 million degrees; in August 1978 the Princeton PLT exceeded 50 million degrees; and the American Alcator has exceeded an nt value of 10^{13}. All these experiments have gained ground, nibbling away by a factor of ten times and again over the years, but still not reaching the thermonuclear breakthrough, which may be expected from General Atomics' Doublet III or the TFTR (Tokamak Fusion Test Reactor) at Princeton currently being constructed in the United States (and expected to start operation in 1982 or 1983). Should these succeed, these experiments will possibly be followed up by experimental thermonuclear reactors prior to commercial prototypes, planned, *at the earliest* — and one knows that nuclear physicists are born optimists — for the beginning of the next century. Only then will it be possible to assess the commercial promise of this new form of energy, which makes it even more intriguing.

"FAST" FUSION WITH LASERS

Work on controlled nuclear fusion apparently started in 1950, more or less tied up with the secret of the hydrogen bomb. After the first Atomic Conferences in Geneva in 1955 and 1958, a certain amount of international collaboration came about in the field of fusion against magnetic containment.[4]

Fusion with lasers made a rather boisterous "civilian" start in 1972 but was then quickly cloaked in an atmosphere of discretion, or secrecy even, because of its military potential. Laser reactors function in fact like hydrogen "microbombs." The enormous energy of high-power bundles of lasers is concentrated on a microscopic drop or blob of deuterium-tritium, which implodes on itself, achieving a phenomenal superdensity. As the implosion tends to be followed by an explosion, the fusion energy required must be liberated over a very, very short time (a billionth of a second, or nanosecond) while this supersolid still exists, before its matter disperses. One counts on the "inertia" of the imploded globule, hence the name "inertial confinement."

Although some spectacular progress seems to have been made, there is today no laser in existence powerful enough to get the projects out of the experimental stage. It remains to be determined what the rhythm of these microexplosions should be and how they should be maintained in order to produce ultimately useful energy. It is a long path.

FUSION OR FISSION? FUSION AND FISSION?

The idea of mixed fusion-fission reactors has been put forward because the large number of neutrons emitted in thermonuclear fusion reactions could be used in a "fertile envelope," producing at the same time tritium from lithium and fissile fuel, plutonium-239 from uranium-238, or uranium-233 from thorium 232. The idea has even been taken as far as to attempt to maximize the production of fissile material, even at the cost of energy production (for economic reasons) in small fusion reactors using lasers, for example, making these in effect true fuel factories.

The advantage would be that one could choose from nuclear fission reactors on the basis of their energy characteristics alone — high-temperature reactors, for example — and be freed of the very difficult problems of optimization entailed in such complex devices as breeder reactors. This would simplify fission by complicating fusion! This would be to fall between Scylla[5] and Charybdis. Fusion purists will comment that this would mean foregoing the advantages fusion has over fission: greater safety, less radioactivity, and no proliferation risk.

In conclusion, as it lies in wait in the shadow of its millions of degrees, thermonuclear fusion may lead to a new energy source, at the same time inexhaustible and safe, but certainly complicated and doubtless expensive, but nevertheless an improvement on nuclear energy; or it could be a fiasco. In the latter case it would be regarded retrospectively as the last fling of some rather dangerous

[4] Under the aegis of the European Economic Community there is a well coordinated program of research into controlled fusion, which has culminated in the common project JET (Joint European Tokamak), which was long bedevilled by a ridiculous quarrel over the choice of site before Culham in the United Kingdom was finally selected.

[5] Incidentally, this is the name of an experimental thermonuclear device.

crackpots trying to impose extremely sophisticated solutions upon a ''terawatt'' universe. In the meantime, the search must go on. Even if it is just to remove all doubt.

4

GETTING INTO HOT WATER

In passing from thermonuclear energy to geothermal energy we return to earth (even into it!), without forgetting to shake, like Cyrano de Bergerac, our cloak of imagination to rid it of the last motes of stardust and artificial suns. No more magnetic fields twisted by superconducting magnets with cryogenic circuits, no more superlasers, but holes, pipes, and pipes.

Indeed, in a book like this on new forms of energy, it is never easy to decide where to put geothermal energy, which has already been used for a very long time (by the Romans, among others). Technologically it should by analogy come after oil, since it uses the same drilling techniques and the same "underground knowledge" (although it is one of those rare forms of energy originally independent of the sun). Technically simple, geothermal energy is on the other hand the least concentrated (at its source) and most diluted of all energy resources: a ton of hot rock contains 1000 to 10,000 times less energy than a ton of coal, which in turn contains 1000 to 10,000 times less energy than a ton of uranium. Geothermal flux at the surface of the earth, about 1.2 microcalories per cm^2 per second, is much weaker than solar flux, which already is pretty thin. The total surface of the earth thus gives off some 30 million MW, or 30 TW (about four times as much as the amount of energy mankind consumes annually) or 5000 times less than the globe receives from the sun. This geothermal flux is so weak that there is no question of using it. But this flux corresponds to a certain thermal gradient, and attempts are being made to domesticate the thermal gradient, which increases the lower one goes below the earth's surface. This gradient is also called the geothermal degree: it is the number of degrees gained given a depth difference of 100 m. In so-called "normal" zones, about 99 percent of the continental area, the geothermal degree is about 3° C per 100 m; in "abnormal" zones that have been explored[1] it can be 5° C, 10° C, or even 20° C per 100 m. (This is still far away from the millions of degrees involved in thermonuclear power.)

Another pecularity of geothermal energy is that the resource itself and the

[1] Explored and feared at the same time, because they are often zones of volcanic activity.

amount of it are difficult to grasp. It is evident that the total amount of heat contained in, say, the first ten km of the earth's crust, above a reference surface temperature of 15° C, represents a considerable quantity of energy and, *a fortiori,* if one considers greater depths. But nobody knows what fraction of all this energy is really accessible, and what fraction is effectively recoverable. Nobody can even imagine this. An interesting consequence is that there are "extreme" world energy models that are entirely nuclear, entirely thermonuclear, or even entirely solar, but we know of no long-term "entirely geothermal" scenarios. (Should one assume that this is because geologists and geothermists are serious-minded people?)

RESOURCES, DEPOSITS, AND TYPES OF EXPLOITATION

There is subterranean heat everywhere. By analogy with mineral or petroleum deposits, places where this heat has accumulated are called geothermal deposits. These are also deposits in the sense that their content, partly fossil, is in general being exploited at a rate much faster than it can be regenerated. There are two different types:

- deposits of dry rock at high temperature, exploitation of which is difficult, but which are without doubt the most numerous and perhaps the most promising in the long term.
- aquiferous reservoirs, which "accumulate" heat, as the caloric capacity of water is nearly double that of rock. J. Goguel has shown that this is not young water expelled from deep-lying rock, but rainwater that has sunk into the ground and is contained in a permeable layer overlying hot rock. It is these deposits we are attempting to exploit today.

Generally one distinguishes, depending on the state of the water (liquid or steam) and its temperature, between high-energy deposits — with temperature generally over 150°-200° C — and low-energy deposits. The high-energy deposits, which correspond to the "abnormal" zones just mentioned, have arisen due to geologic phenomena that are not well understood but probably recent (between a few to several tens of millennia) and, linked to volcanic activity, are limited in depth (for example, a granite massif formation).

The different types of geothermal exploitation due to these differences are:

1. Dry steam deposits. These are the most interesting, but the most rare. The steam may be transmitted directly to a turbine to produce electricity, probably the cheapest form of energy it is possible to produce. Currently only five deposits of this type are known: Larderello in Italy, the geysers in California, Valle Caldera in Mexico, and Matsukawa and Otaka in Japan.
2. Hot water (or damp steam) deposits are much more numerous, including the major volcanic zones: in Alaska, Hawaii, Chile, Peru, San Salvador, Mexico,

and various regions of the United States, Guadeloupe and Martinique, Iceland, numerous regions in Europe (France, Italy, Central Europe, Greece,); Turkey, the Caucasus, India, Ethiopia, Djibouti, Kenya, Reunion Island, Indonesia, the Philippines, China, Australia and New Zealand, Kamchatka, and of course Japan, most of these countries being in the volcanic belt.

Hot water can be used directly for industrial heating or indirectly to produce electricity by one of two processes: by expansion in the well and the production of damp steam, which must be dried by separation before being transmitted to the turbine, or by using binary cycles with propane, isobutane, freon, or ammonia.[2]

3. Warm water deposits, as we continue to descend the temperature scale, under 150° or even 100° C. These exist in most sedimentary areas at 1000 or 1500 m down, or lower. In France, for instance, they represent the major part of geothermal resources. One can if necessary still produce electricity by having recourse to binary cycles, but such production currently appears very marginal in economic terms (a plant near Peking, of about 100 kW, is said to function using water at 86° C and ethylene chloride). The most interesting use is for industrial and domestic heating. This is without doubt the most important outlet, and the most immediate, for a low-energy geothermal resource, providing that it is close to cities or highly populated areas (so the Paris basin would be favorable, but the Aquitaine basin, in the Southwest of France, would not).
4. Geothermal energy from dry rocks. This is still in an experimental stage in the United States and elsewhere (at the Los Alamos Nuclear Laboratory in New Mexico). The idea is to exploit a hot geologic anomaly by drilling a network of producing and injecting wells. Having fractured the rock between the wells (by hydraulic methods or, as thought at one time, by nuclear explosions), a fluid would be injected, probably water to be heated and/or vaporized. To the drilling cost must be added the cost and risk of fracturing the rock: this might result in a heterogeneous network of irregular fractures, with circulation limited to only a few passages, with the risk of rapid cooling. But the idea is interesting and worth pursuing. It would lead to a search for the necessary energy in the bowels of the earth, with temperature increasing with depth. The cost of drilling, always a handicap in geothermal prospecting, would here be of the utmost importance, for in general the cost of drilling multiplies by a factor of 2.5 from 750 to 1500 m and by 4 again from 1500 to 3000 m.
5. The dreams of scientists and others. Numerous scientists have been attracted by the idea of someday utilizing the enormous energy of volcanos, or magma. The eruptions of Bezymiamy in Kamchatka, for instance, represented an

[2] "Binary cycles" use a different fluid from steam to drive a turbine, this fluid having a higher pressure for the same temperature, which has the advantage, among others, of reducing the size of installation. The cycles are called "binary" because they use two fluids (hot water and a "driving" fluid).

energy of 40,000 billion kW.h, with some 2.5 billion tons of rock pulverized, these having an initial velocity of the order of 2000 km/h. Who is not tempted at the thought of such power? But it is a long way from the mouth of the volcano to energy utilization, and no technically feasible project is known.

On the other hand, the Sandia Laboratories in the United States have suggested creating a new source of hydrogen by getting water to react with the mineral ores constituting the magma; if cellulose waste were added, one could even produce carbon monoxide and methane. But this presupposes enormous progress, even a revolution, in drilling techniques; it would be better not to count on these in the short term.

UTILIZATION OF WARM WATER: THE FRENCH EXAMPLE

There are a number of programs planned or even realized using warm water (say, under 200° C) in New Zealand, Hungary (in the Szeged region), Iceland, and other places. Following the 1973-74 oil crisis, France, desirous of turning every usable domestic resource into heat, launched a very bold program of low-level geothermal energy (the only one apparently available). It did not start from scratch: two props for this important program were foreknowledge of potential resources thanks to knowledge acquired from the feverish search for oil after the war (with some 5,000 drillings) and the pay-off from a pilot operation at Melun-l'Almont since 1969.

As regards resources and recoverable reserves, the following considerations well illustrate their difference in nature and degree. Basic resources are immense (the first 10 km with temperatures higher than 15° C, over the area of France) being equivalent to some 20,000 billion tons of oil, which means nothing. The main priority assigned to the BRGM (Bureau de Recherches Géologiques et Minières) and the DGAST (Délégation Générale à la Recherche Scientifique) was to determine the possible quantity of accessible resources real or realistic, then of *reserves* in situ, and finally, *economic and technically recoverable reserves*. They wisely started by making an inventory of the aquifers, large anomalies in temperature being small and rare.

According to J. Lavigne of the BRGM, resources identified in France are estimated at 380 $\times$ 10^{18} calories, the equivalent of 38 billion tons of oil,[3] principally in the sedimentary basins of Aquitaine (300 $\times$ 10^{18} calories) and the Paris region (67 $\times$ 10^{18} calories), followed by Alsace, Bresse, the Rhone valley, the Mediterranean Midi, and Limagne. As regards in situ reserves the positions of Aquitaine and the Paris basin are reversed, clearly illustrating the importance of the economic factor in the concept of reserves. In situ reserves have thus been put at 10.8 $\times$ 10^{18} calories, mainly concentrated in the Paris area (7.2 $\times$ 10^{18} calories). Only a fraction of these in situ reserves is thought to be recoverable currently, both

[3] 10^{10} calories = 1 TOE (ton oil equivalent)

from the economic and the technological point of view, representing 1.8×10^{18} calories, equivalent to 180 million tons of oil, that is to say, permitting a production of at most 6 M TOE annually for thirty years.

It is interesting to look at the Paris basin project in some greater detail, a choice target for geothermy because of the size of the population and its urban habitat, and because of the existence of numerous aquifer layers on different geological levels: Lusitanian, Lias, Trias, and especially Dogger.

The surface area over which the Dogger (and primarily its upper reservoir) has temperatures over 50° C (average temperature 60° C), a thickness of over 50 m (average thickness 100 m), and good permeability is put at 15,000 km^2. In situ geothermal reserves per km^2 have been estimated at 3×10^{15} calories, equivalent to 300,000 TOE/km^2. Taking into account the fact that within the area Compiègne-Meulan-Melun-Epernay, which covers 8000 km^2, about a quarter of the surface area is urban or may become so, which is to say the geothermal exploitation is possible there, one gets 6×10^{18} calories, the equivalent of 600 M TOE for in situ reserves. About a sixth would be recoverable, the equivalent of about 100 million tons of oil, or production of 3.3 M TOE annually for thirty years, which is no negligible quantity. Recovery of reserves could be radically improved by rationalization of management of all the water formations.

It is the Dogger formation where geothermal exploration has already taken place,[4] such as at:

- Melun-l'Almont (depth 1800 m, temperature 70° C): hot water and heating for 3000 homes
- Villeneuve-la-Garenne (depth 1800 m, temperature 58° C): for 1800 homes (a pioneering effort by the Total oil group)
- Creil (depth 1650 m, temperature 57° C): an important project for 4000 homes, the first large-scale combined use of geothermal energy and heat pumps
- Melun-Mée-sur-Seine, in construction for 6000 homes
- Meudon-la-Forêt, planned for 7000 homes.

All these projects use the "doublet" principle, that is to say, two wells with their heads close together but with their bottoms 800 to 1000 m apart, a well for production (artesian, or with pumping) and a well for the re-injection of brackish cooled water. This re-injection which minimizes pollution problems, is of course expensive, requiring a second well, but it has a beneficial effect on the water table, maintaining the pressure and optimizing the output of the wells. The high salinity (5g/1tr) and the corrosiveness of the brackish Dogger water means that some special materials have to be used for the piping (glass fibre at Villeneuve-la-Garenne, for example) or for the heat exchangers (made of titanium, for example).

Projects carried out to date permit a saving of about 8000-9000 tons of fuel per

[4] Also active is Mont-de-Marsan in Aquitaine, with one well in production (run by the oil company Elf-Erap).

year, an amount that will be approximately doubled when projects under construction or in the planning stage come into operation. In fact, following the oil crisis of 1973-74 the government in 1974 launched a program of incentives. The investment necessary, initially heavier than for normal heating systems, requires new methods of financing. This is why the Géochaleur Company was created, whose object is to investigate the market for homes with geothermal potential and to help with setting up the financial and technical side of new operations. In the meantime, the amount of financial aid has gone from 7 million francs ($1.7 million) in 1975 to 37.5 million francs ($9 million) in 1978. Finally, the Act of 19 July 1977 is concerned with existing heating contracts, which must not become an obstacle to the application of new forms of energy.

Things have undeniably got started, but the start was slower and more difficult than foreseen, because of institutional problems, among others. The initial objective, admittedly ambitious, of 500,000 ''geothermal'' homes in 1985 will probably not be achieved. Another reason for this is that the housing program itself has been cut back (from a target of 450,000 homes per year) and is also changing in structure: fewer enormous multi-family blocks and large units, more blocks for smaller numbers of families and individual homes, for which the geothermal payoff is lower. Currently, the target is 800,000 to 1 million ''geothermal'' homes from now to the end of the century. With a doublet supplying on average 2000 homes, this represents 500 doublets, which is certainly attainable. Taking 3000 TOE annually for a doublet bore-hole, costing about 8 million francs, or nearly $2 million (1977 prices), this program represents a saving of about 1.5 million tons of oil per year (about 1 percent of France's current energy consumption) for a drilling investment of 4 billion francs (near $1 billion).

WAITING MODESTLY

Although by no means negligible, these figures suggest a relatively modest role to be played by geothermy. (From now until 2000 high and low-energy sources could contribute 1 or 2 percent of the world's energy supply.) This share could probably be increased by research aimed at electricity production (this being easily transportable) using, for instance, binary cycles at low temperatures.

Currently, on a world scale, geothermal electricity in dry steam deposits represents a capacity of about 1200 MW(e), the equivalent of *one single* nuclear reactor. More than half this amount of 1200 MW(e) comes from the geysers in California, and a quarter from Lardarello. The rest is divided up between Japan, Mexico, New Zealand, and the USSR.

While waiting for a possible breakthrough in the thermal exploitation of dry rocks, which would provide abundant good-quality resources but ones inevitably very diluted and expensive, it is perhaps necessary to put the geothermal problem into the wider frame of, on the one hand, low-temperature energy requirements (more than 50 percent of final energy requirements) and, on the other, a complex of different low-temperature resources, usually considered individually: geothermal aquifers, waste hot water from thermal power stations (an enormous "deposit," as two-thirds of the calories produced are thus "wasted") and possibilities of solar heating. Certain couplings have been suggested or investigated: the coupling of nuclear and geothermal power, and of solar and geothermal power, the latter providing in each case a means of large-scale energy storage, storage being a problem with nuclear and solar energy. A lot remains to be done to aid understanding and improve management of this vast domain of hot water.

5

THE THOUSAND AND ONE POSSIBILITIES OF SOLAR ENERGY: FOR BASIC NEEDS OR JUST FOR TOPPING UP?

In our opinion there is no question as to whether solar energy will ''eventually'' be developed. After many hopes and many false starts, the process *has* got off the ground.[1]

Where, in what country, will solar energy be developed? In countries rich in economic and technological respects — the sun gobbles up money and technology — but relatively poorly off for sunlight or ''good'' sun? Or in the developing countries, poor in monetary terms and technology — both money and technology having so many calls on them — but rich in sun?

When and with what rhythm will solar energy develop? As a ''catastrophe'' or ''crash-program'' that some people are trying to impose (whom we shall refrain from calling ''solar fascists'', however) or, as others wish, hoping that the sun will be able to cover an already appreciable part of our needs by the end of the century and nearly all of them in fifty or a hundred years? Or, on the contrary, taking over slowly, from a base of economic competitiveness and following the free play of market forces, which could take a century or, worse, till the Greek kalends, if one can wait that long. But perhaps it can be cunningly steered between these extremes? And how? In the form of heat, electricity, hydrogen, the ''petrol tree''? Will it be a basic form of energy arriving in time to meet all our energy requirements, this being appreciated from the outset, or just energy for topping up, allowed to coexist with other resources, with society seeking to or having to make do with whatever it can.

It is not easy to reply to these questions. In my opinion it is not even desirable to reply to them or, at any rate, not too quickly. It seems evident that a process of ''solarization'' must be an act of will, which is not a synonym for a dictatorial act,

[1] It could have been said, not without some irony, that ''literature about solar energy is developing faster than the realization of solar plans, but the goals are in sight.'' Indeed it looks as if 1978 was the year in which solar energy took off.

to have a chance of penetration within a reasonable time (or by when it is necessary), given that market forces are blind and/or too slow. This will, this decision only makes sense within a framework of intense reflection and of a wish to understand the mechanisms of our society, which is desirable not only with an eye to energy. If this understanding is sufficiently profound, which is to be hoped will be the case, it should lead to a range of choices. The process will be slow. It will take time, and we do not know whether we still have this time. It is not sufficiently realized that this is a new situation. It is not a pleiad of thinkers or a handful of revolutionaries who are called upon to reflect on the future of mankind. It is man himself, society, the nations. And here time, not energy, is our more precious resource. It is interesting that both nuclear and solar power have contributed to this jab at our conscience, to initiating this process, at the same time exalting and dangerous. We shall come back to it.

SOLAR ENERGY: THE RESOURCE

The basic attraction of solar energy is its vastness on the planetary scale and its perennial quality as regards the time-scale. Tidal energy, for example, is also vast, viewed on the scale of billions of years on earth; but on a global scale, and in the light of our annual energy requirements, its power is relatively modest.

Almost eternal, viewed on a human scale, the solar energy received by the earth is equally considerable. In the upper layers of the atmosphere — where, incidentally, certain projects have envisaged capturing it[2] — the solar flux is about 1.35 kW(th)/m^2. The terrestrial "disc" intercepts 178,000 TW (10^{12}) or 20,000 times more than the 8 or 9 TW that mankind uses. (In 1977 we consumed about 8 million TCE, approximately equal to 8 TW-years, an annual quantity of energy corresponding to permanent utilization of a fictitious instantaneous power of 8 TW.) In practice only a fraction of this energy reaches the earth's surface, for the atmosphere acts as a reflecting and absorbent medium, diffusing, turbulent, and irregular, if not unpredictable. This still considerable fraction of solar energy is responsible for our climate, and is essential for our food and for many other things as well.

It is convenient here, as we are talking of resources, to make a preliminary remark. A McKelvey diagram is used to depict energy stocks of oil, gas, coal, and uranium, this being useful to distinguish between economically recoverable reserves, identified but uneconomic resources, and resources yet to be discovered.[3] For solar energy, energy-flux, this does not apply. There is no comparable diagram. One could of course distinguish between the fraction that can be economically produced under present-day conditions or under conditions foreseeable in the short term, leaving the fraction yet to be tamed almost infinite

[2] These solar satellite projects prove — as if it were necessary — that the atomic scientists do not have a lien on sophistication. Remember too the projects for damming the Mediterranean.

[3] See Chapter 7 on oil.

compared to the scale of our needs. But both are renewable, inexhaustible. This is an essential difference.

The solar flux reaching the earth's surface has the property — the failing, say its detractors, the inconvenience, say its proponents — of constantly varying in space and time. The variation in terms of space can be seen from Table 5-1.

TABLE 5-1. Some Solar Data

AREA	AVERAGE ENERGY kW.h (th)/m²-day	AVERAGE POWER W/m²
Deserts, tropics	5 - 6	210 - 250
Temperate zones	3 - 5	130 - 210
North-Eastern Europe	2 - 3	80 - 130

One may be struck by the fact that there is only a factor of 2 or 2.5 of difference between the sunniest regions and western Europe, an area not thought to be overblessed with sun. Though not a latent defect, this is nevertheless an important factor in economic terms, as we shall see: it introduces a difference of both quality and cost (rather like oil, with massive or medium deposits).

As for time, the amount of sun naturally varies with the seasons, by the day, by the hour, in a regular and predictable fashion, and from one moment to the next in nebulous, irregular, and unpredictable fashion.

The question has sometimes been asked, what would have become of our society after the war without the discovery of Middle-East oil? That is, how it would have fared without access to a source of energy so abundant that it seemed inexhaustible and so cheap that its relative cost went down for at least fifteen years. That situation, it cannot be repeated enough, was totally abnormal; we thought it permanent, however, and thought we could reproduce it or bring it about again with atomic energy at the end of the 1950s. The reply most often given is that we would have progressively exploited more expensive sources of energy and that they would have been used in a much more economical way. Europe, among other consumers, would have continued to produce and use coal, coal at least as dear as oil today. We would also have sought, more intensively, no doubt, dearer oil, as has always been the case in the United States. No doubt we would have sought more aggressively to make use of resources that were abundant but difficult, such as the bituminous shales of the Paris basin and elsewhere. In short, it may be imagined that our economies, *because of their internal dynamism,* would have developed in just the same way, more stingy with their wealth, more "European," say, than the Europe of today.

Let us take the question much further. After all, so much effort is being devoted nowadays to imagining the future that we might well spare a moment to imagine a different past. How would our society have evolved *without* fossil fuels: Would it have remained an agricultural society, or would it have developed solar energy? Remember the windmills of Holland and Denmark; remember the hydraulic energy that permitted the first flight of electricity, the machines of Lavoisier and Mouchet, the fermentation of alcohol, charcoal, etc.[4]

We will not pursue this imaginary exploration further, but just note that "the road not taken," to pick up a profitable expression dear to the heart of Lovins, is in a sense the one proposed to us today. We have the benefit of a technological tool that is infinitely more powerful (but also more expensive) than existed a few centuries ago, but we have the disadvantage of having acquired bad habits (knowing these are the hardest to get rid of!) of wasting energy, of shifting our rhythms of activity more and more from the solar rhythm that used to guide our work and our days.

The ease of storage, that is, the availability of energy anywhere at any time, finds its culmination in oil. The advantages of conversion[5] and flexibility of use find their culmination in electricity. Because of our bad habits we would like to have solar energy with the advantages of oil and electricity. Is this within the realm of possibility? If so, at what price? We shall find out by looking at the ways in which solar energy can be harnessed.

THE HARNESSING OF SOLAR ENERGY

One could almost reply straight away, intuitively, that the methods of harnessing solar energy are so numerous and so diversified that there must be one which meets our requirements. We think so ourselves, as the possible combinations of solar energy and our technological resources are practically infinite. But curiously enough, this multiplicity of solar power, so attractive and promising in the long term, does it a disservice in the short term. There is no golden road today on which all our efforts should be concentrated or, within some particular energy system (solar cells, for example), some variant more promising than all the others. Their characteristics and their promise invite us to explore them all. Their cost, the time needed for experimentation and for comprehension constrain the adoption of a slow and measured pace, which also applies to market penetration. The diversification of needs and the means to satisfy them hinders every royal road. (There too, the abundance of energy resources has led to an abundance of alternatives.)

Put broadly, three principal paths are being followed in the attempt to harness solar energy:

- in thermal form: the sun's rays are captured with the aid of collectors (or catchers) and are used either for heating or cooling buildings or for industrial or agricultural purposes (more than 70 percent of energy is used in the form of domestic or industrial heating, most of it under 150° C).

[4] We think the principal force of Europe was its internal dynamism, its thirst for progress, its will to develop, rather than the abundance of coal — without denying that this made things much easier.

[5] The conversion of primary energy into secondary energy such as electricity, with more flexibility in use.

- production of solar electricity, thermodynamically or with photocells. Hydroelectricity, the utilization of the oceans' energy and the conversion of wind power to electricity also come into this category.
- biological conversion of solar energy, that is to say, production of fuel from fast-growing vegetables, agricultural waste, or other organic products. These fuels can be used as such or to produce electricity.

Thermal Applications

Solar applications are termed "active" or "passive," depending on the means of utilizing solar energy.[6] Before reviewing the various "active" means of using solar heat, let us recall that solar architecture offers a vast potential for improving homes currently designed like thermal sieves. These "passive" means make use of characteristics of the building (openings, materials, and so on), using the external climate to create a favorable climate inside. The role of the heating system is thus minimized (heating requirements can be reduced by 50 percent or more). The utilization of passive means should always precede that of active systems.

Water heating is the most common use (in the United States, Japan, Israel, and other countries) and the most promising application for solar energy in the short term. This relative commercial success has several reasons: the first is that the need for domestic water is relatively constant across the year, thus assuring a continuous amortization of the investment; the second is that the most common solar water-heating systems can very often be connected up with an existing hot-water system. The installation has a moderate cost, comprising a collector (\$150-250 per m^2) (see Figure 5-1), a storage unit, a heat exchanger and a pump; and it also generally incorporates a back-up heating system (fuel oil, gas, or electricity). The yield from the collectors is generally quite good at the temperature involved (about 60° C), which means that such an installation can be used even in areas with little sun (efficiency 30-40 percent). For instance, the Délégation aux Energies Nouvelles has set a target of 3 million solar water-heating systems in France in 1985, which shows solar potential in a European country with moderate solar characteristics.

Heating buildings is currently being strongly encouraged in the industrialized world because of the importance of heating within the energy budget (30 percent of expenditure on energy in general), but it is somewhat more difficult and a little less attractive. The first objection is the gap between the seasons when the sun's rays are most intense (May to September) and the period when heating is required (October to April in a large part of Europe). Hence the limited role of partial heating, leading to a doubling of some of the installations. The collector area is generally determined with an eye to winter solar radiation, thus losing some of the summer radiation. It cannot be adapted to existing buildings, except with difficulty. As for new buildings, it is not so easy to lay down ideal norms for

[6] "Active" are those that require equipment or installation, while "passive" means relate to the design of the building itself.

insulation, and even less easy to get these rapidly accepted by the entire building profession.

The integration of the collectors into the architecture also presents some problems. The installation comprises collectors that use water or air (as distinct from water-heating systems, which use only collectors with water), a heat exchanger, a distribution system for the heat, a storage unit (using water, sand, stones, or special liquids, organic or mineral, or eutectics[7]), a control unit, and a back-up system. The heat pump appears particularly interesting as a back-up though its economic advantage has yet to be demonstrated.

These active systems can meet up to 50 or 70 percent of the total thermal load with an annual utilization factor between 35 and 20 percent. Compared with the current cost of fuel oil, which will not decrease, the period of amortization for the additional cost of installation can vary between ten and twenty years.

Seasonal storage over several months does not seem feasible for individual homes using currently available techniques. The volume of the storage unit would need to be nearly as large as the home to be heated. The situation is more favorable for blocks of dwellings (where one must reckon on about $1 per m^3 of storage for each $5 of installed collector area). An interesting solution currently being studied would be to link the system to an artificial geothermal reservoir.

There are also numerous experiments being carried out with "self-sufficient" houses, with solar heating, electricity from wind power, a south-facing greenhouse, semi-underground, or so-called bioclimatic houses (with walls that might consist of membranes with variable characteristics). In Denmark there is a "zero energy" experimental house (42 m^2 of flat collectors, seasonal storage of 30m^3 of water, and a central greenhouse). The ideas are interesting, but the results in the first year were rather disappointing: the solar system met only 43 percent of the thermal requirements.

Cooling of buildings is much less in demand but has two big advantages: the sun is at the top of its trajectory at the time when cooling is most necessary, and the electric networks are generally working with surplus capacity at the hottest periods. The systems that have received most attention use absorption refrigerators, which unfortunately require higher working temperatures than are required for domestic hot water. At present there are no inexpensive collectors on the market with adequate efficiency at temperatures around 90° C. These problems, together with the same ones that apply to solar heating (difficulties of definition and the acceptance of norms), suggest that the spread of solar cooling will be slower than that of solar heating.

Various other thermal applications being developed look like having a promising future (though their contribution to the energy balance will probably be modest), such as the drying of harvests, the desalination of sea-water or of brackish water, solar ovens with parabolic mirrors for high temperatures (such as the

[7] A mixture of solids and liquids with high calorific capacity, that uses the latent heat from changes of state (solid → liquid, or liquid → solid)

FIGURE 5-1. (A Flat Plate Collector)

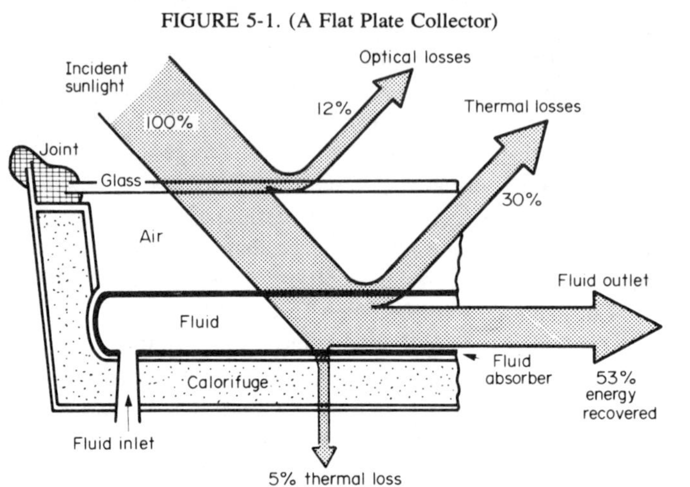

A flat plate collector without concentrating elements is composed of three elements:

- an absorber, in which water circulates. The surface exposed to the sun is designed to absorb the solar rays and has a coating either just of matt black paint or a selective layer characterized by good absorbing power for the long waves of the solar spectrum (0.3 micrometres, or 0.3 μ) and a low emitting power for wavelengths corresponding to low-temperature radiation from the absorber (5 to 12 μ) (the "greenhouse effect").

  This absorber can be directly derived from a commercial heating radiator, made from tubes welded onto a copper or stainless steel plate, or made from aluminium with the rolled and bonded technique.
- an insulating case made, for example, from polyester and glass fibre, protected by a gel coat, and with a polyurethane layer 45 mm thick, giving a low coefficient of heat transmission from the rear surface. The casing can also be of steel plate or aluminum, with glass wool insulation.
- a transparent cover: horticultural glass 4 mm thick, special glass, nonreflecting (more expensive); material resistant to ultraviolet rays.

A good flat plate collector inclined at an angle of 30° can in Central Europe provide between 850 and 1300 therms per m^3 per year, according to the region, or the equivalent of 85 to 130 kg (100 to 14 ltr) of fuel oil.

pioneering Odeillo solar oven, in operation since 1970, producing temperatures of over 3000° C, with a concentration factor of over 10,000), and so on. Mention should also be made of solar pumps, where the leadership is held by a French company, SOFRETES. These pumps work with flat collectors (the only ones capable of working equally well with diffuse energy, in cloudy weather) and a thermodynamic cycle using butane or freon. There are sixty or seventy pumps, with varying power, currently in the world. To take an example, an installation in the Sahel with 80 m^2 collector area pumping 5 m^3 per hour, five hours a day with a shallow water table (about 30 m underground), can satisfy the water requirements of a population of 1,250 persons or water 625 head of cattle (40 litres per day per head) or irrigate half a hectare (0.6 litres per second per hectare).

Of course, installations similar to those used for domestic water heating can also be used to produce low-temperature industrial heat, although the most interesting and most usual temperature range is from 80° to 250° C, which requires a certain concentration of solar radiation. For higher temperatures it would be necessary to have a higher concentration of rays. These devices could be used to obtain high temperatures or, via these, to produce electricity.

Production of Electricity

The possibilities for electricity are numerous. Some applications are already age-old, such as hydroelectricity and wind energy (two forms of solar energy), while others, more recent, are the subject of R & D programs, such as ''solar-thermal electric'' conversion of the thermal energy of the oceans and photovoltaic energy.

A solar thermal electric conversion system (STEC) generally consists of an optical concentrator, an absorber for the thermal energy collected, a heat transfer system, and a conventional part (such as turbine, condenser, pipelines). Leaving aside the conventional part, it may be said that none of the first three systems is really new or very sophisticated (as compared to thermonuclear systems for example), but neither is any of them currently available on the market. Experiments still have to be made to optimize them, to integrate them, to produce them on a commercial scale.

A very important parameter that must be defined is the concentration factor *C*, the ratio between the surface receiving the solar flux and the surface of the thermal collector receiving the concentrated radiation. It more of less determines the type of plant, and can vary between 1 (flat plate collectors with solar pumps) to more than 10,000 (Odeillo). It can be obtained using lenses (Lavoisier's machine of 1746 for achieving high temperatures, a solution abandoned today for electro-solar power) or by reflecting mirrors (as in Archimedes' mirrors at Syracuse and Mouchot's machine of 1861 in the Tuileries). Cylindrical-parabolic concentrators give concentrations of about 50, permitting temperatures around 200°C to be reached in channels placed at the focal point. Spherical or parabolic reflectors give concentration factors of about 300 and are suitable for small power plants of less than 1 MW(e) using a number of collectors. Systems of mirrors reflecting the sunlight onto a boiler at the top of a tower permit still higher factors of concentration, of the order of 1000. Not requiring complex and very precise reflecting surfaces, this concept is universally recognized as the best solution for high power requirements of above 1 MW(e). But the concentration generally means a control system is necessary for the optical systems, so that their orientation can be varied according to the position of the energy source, the sun.

The exit temperature of the fluid from the boiler, which determines thermodynamic efficiency, depends on the concentration factor *C* and also on the flow rate of thermal fluid and thermal losses from the boiler (which limit the maximum temperature that can be obtained). These losses are partly due to the reflected radiation from the walls (infrared emission, a function increasing with temperature) and convection losses. To optimize the boiler, especially in big installations, it is designed as an almost closed cavity (to reduce losses from reflection and radiation), open at the bottom. The actual efficiency of the installation takes into account the thermal losses of the boiler and the other systems and the optical losses from the concentration systems. For past installations and current ones the total efficiency will be limited to between 5 percent and 20 percent, in time it is hoped to come closer to the theoretical figure of the order of 40 percent.

For small power plants with distributed collectors it is necessary to seek a compromise between the efficiency and the cost of the concentrators. The optimal solution is not evident, and perhaps not unique.

One can use a fixed spherical collector (partly buried), focusing on a boiler mobile in two directions. For example there is the Pericles project of the Marseilles astronomy laboratory, with a power of 40 kW and a sphere diameter of 40 m. In this concept the number of mobile mirrors is minimized, but not all the reflecting surface is used all the time.

Another solution (that of the Marseilles heliophysics laboratory) uses a parabolic heliostat with an area of 50 m^2 pointed towards the sun, with a boiler and organic liquid storage. These parabolic machines lend themselves to medium-power installations by juxtaposition of modules, each feeding a central thermal plant through a network of pipes. But the thermal and hydraulic losses (pumping) limit this type of installation to a few hundred kW.

A boiler at the top of a solar tower (100 to 250 m) receives converging images of the sun from a large number of mirrors or heliostats, which follow the daily movement of the sun. These heliostats are of course the critical part of the system. They are flat mirrors (glass with a silvered rear face, or possibly aluminum) or, better, slightly focusing (that is, with a very slight curvature). The reflecting surface has to have a good coefficient of reflection and must maintain it over time (a possible maintenance problem about which little is known at present). Each heliostat, with an area of 30 to 60 m^2, can be oriented around two axes and is mounted on a structure the rigidity of which is such that the accuracy of aiming is within a few milliradians up to a certain limiting wind speed. Another solution is to protect the heliostats by plastic domes. The problem, which is far from simple, is to minimize the cost of the heliostats for a given performance.

''Aiming'' toward the sun is controlled automatically (far from the ''agricultural mechanics'' that some people regard solar systems as having), or by individual computer control using an aiming device (closed loop), or by centralized computer control unit fed with the coordinates calculated for the sun (open loop).

The positioning of the heliostats has to minimize shadow. The total site area may be 2.5 to 4 times that of the spaced mirrors (currently the non-exclusive utilization

THE ODEILLO ELECTRO-SOLAR PLANT

In November 1976 at the laboratory of solar energy of Odeillo, France, a thermal-electric solar power plant supplied electricity for the first time to the French electricity network. In one sense this was also a world premiere.

The field of mirrors belonging to the big solar furnace was used, with at their focus a receiving boiler containing as thermal fluid an organic liquid (hydrogenated terphenyl) heated to 335° C by solar radiation. Using a storage tank of 30 m^3 and three heat exchangers, steam was produced at 270° C and 27 bars, this driving a turbo-alternator. The average power produced was 64 kW, which means the overall conversion factor was around 8 percent.

This experiment demonstrated that a solar thermal electric solution is feasible. Its main aim was not to give some idea of the possible future cost of solar electricity, because the Odeillo collector system is very different from the fields of heliostats currently envisaged for future power plants.

of the site area is being studied). The size of the field of mirrors and the number of heliostats cannot increase indefinitely without the quality requirements for the precision of the mirrors and the aiming system becoming prohibitively expensive.

There is general agreement today that the number of heliostats should be of the order of 1500 to 3000. For instance, taking 1500 heliostats of 50 m^2 would give modules of about 10 MW(e), with the possibility of thermal coupling of some ten of these modules to obtain plants of around 100 MW(e), compared with 600-800 MW(e) or even 1300 MW(e) from nuclear power plants.

After the problem of heliostats, the other big problem is that of storage, since the plants would in principle operate only six to eight hours a day, corresponding to the hours of optimum sunlight. The solar thermal electric method incidentally has the advantage with respect to the photovoltaic system of permitting thermal storage, relatively simply but unfortunately expensively. According to requirements, which are not yet completely clear at present, short-term storage for a few hours, storage for a day, or for several days may be envisaged. The admissible investment may be relatively large in the first two cases because of the greater frequency of loading and unloading.

The cost of heliostats is estimated at around \$250/m^2, with the hope of seeing it come down possibly to about \$100/m^2, which could only be achieved (and verified!) by mass production. A total cost of \$2000/kW(e) seems feasible within a few years from now, which would make the cost per kW.h about 15 cents, a sunny figure.

Table 5-2 shows that numerous projects are in progress all over the world proving, if proof be necessary, that electro-solar power has taken off. These

Table 5-2. Solar Tower Projects Around the World

Name of project	Power (kW(e)) and year of starting operation	Location	Organizations responsible	Optical system	Solar receptor characteristics	Heat storage data	Thermodynamic cycle
SSPS-A International Energy Agency	500	Almeria (Spain)	I.E.A. DFVLR	150 mirrors each $40m^2$	— semi-cavity — opening: 3m × 3m — fluid: sodium — t = 575° C	— 4 hours — fluid: sodium 475° C	Superheated steam at 500° C P = 100 bars
CESA 1 (Spain)	1,000	Almeria (Spain)	Centro de etudios de la Energia	250 mirrors each $36m^2$ S = $9{,}900m^2$	— cavity — fluid: steam 130 bars — t = 520°C	— 3 hours — fluid: melted salts at 340° C	— Superheated steam at 500° C 100 bars (solar operating phase) — 330°C 15 bars (storage phase)
Japanese project	1,000	Shikoku Island	Mitsubishi	850 mirrors each $16m^2$ S = $13{,}600m^2$	— cavity — superheated steam 400° C 40 bars	— 4 hours — 4000 kWh	— steam
European project	1,000 1980	Sicily	CEE	146 mirrors each $23m^2$ + 70 mirrors each $50m^2$ (CETHEL) S = $6{,}860m^2$	— cavity — opening 4.5 × 4.5 m — steam t = 510°C P = 64 bars	— ½ hour at reduced power — fluid: steam and melted salts	— Superheated steam 510°C 64 bars
Thémis France	2,000 (nominal) 3,000 (peak) 1981	Targasonne (Near East)	CNRS-EDF	350 mirrors each $50m^2$ S = $17{,}500m^2$	— cavity — opening 4m × 4m — flange: 6.50 m — melted salts 450° C	— 6 hours — fluid: melted salts 425°C	— Superheated steam 410°C 50 bars
Pilot plant, USA	10,000 1981	Barstow California	DOE McDonnell-Douglas	1760 mirrors each $38m^2$ S = $66{,}880m^2$	— external receptor — steam 516°C 104 bars	— 3 hours — organic fluid and stones	— Superheated steam 510°C 100 bars (solar operating phase)

projects are without doubt far from being commercial, but less far perhaps than the first experimental nuclear reactors were.

Among the problems remaining to be solved, and progress to be made, we may note the reliability and cost of the heliostats, automatic control of the fields of heliostats, the radiation boilers with selective surfaces, the management and technology of storage systems, integration into the electrical grid (except for isolated low-power stations for the developing countries), and promising research into high-temperature thermodynamic cycles with high efficiency (concentration factor of 3000, temperature 1000° C, gas turbines).

The idea of using sea water from the ocean surface as a hot source and deep water as a cold source was first proposed by the French: d'Arsonval (1881) and Georges Claude (1929, 22 kW steam turbine in Cuba). The temperature difference under optimum conditions being 10° to 20° C, the efficiency is at best about 1 to 2 percent. Despite such poor efficiency, it has been calculated that a network of stations spaced 15 km apart over a belt of ± 20 degrees of latitude would represent a potential capacity of 50 TW(e), a sufficiently significant qualtity to justify a fresh look into the question. Compared with solar thermal electric systems, ocean plants using the thermal gradient have the advantage of permitting large capacities of several hundred MW(e) and having thermal storage capability at hand. They have the disadvantage of their low efficiency, which results in ''monstrous'' installations. If they are installed far from the coast, at the best spots (within the tropical ocean belt) they represent transport problems as regards the energy produced (by hydrogen?) or on-site utilization (as for ammonia or aluminium production). Some United States projects by TRW and Lockheed have been judged viable, and currently form part of an experimental programme, with a 5 MW(e) pilot plant planned for 1980 and an industrial-scale unit of 100 MW(e) for 1983.

As an example, the Lockheed project (160 MW(e) net) would comprise a central platform (60 m in diameter and 130 m high), four mobile modules each containing a complete set of apparatus for the cycle (evaporator, condenser, turbo-generators, ammonia pumps — low pressure steam — and sea-water pumps), a telescopic pipe in concrete to catch cold water 470 m down, and an anchoring system. The whole would weigh some 360,000 tons, including 7000 tons of titanium for the exchangers. Installed less than 60 km from the coast in the Gulf of Mexico, for example, such a device would pass on the energy produced in the form of electricity via a submarine cable transmitting direct current. At the equator and far from the coast, it has been calculated that the hydrogen produced would cost about $100 per barrel of oil equivalent.

Such a project seems to belong to the realms of science fiction. But the oil platform made of concrete invented for the Ekofisk deposits in the North Sea might have seemed to fall into the same category twenty years ago.

The ''photovoltaic effect'' involves a difference of potential appearing in an illuminated solid material, usually a semiconductor. Photons (elementary particles of light) interact with matter and lead to the appearance of electrons and positive holes. The movement of these charged carriers produces an electric current, which can give an electric charge.

In practice, the photovoltaic approach entails the direct transformation of solar radiation into electricity in a semiconductor material suitably prepared. It is well beyond the demonstration stage, and it has been widely used in space programs, though there is still scope for progress, which could rapidly improve its prospects. In a word, its promises and its challenges are formidable. It is perhaps the (elegant) solution to the energy problem in the long term.

The advantages of the photovoltaic approach are that it permits the possibility of power plants without moving mechanical parts and hence with a long lifetime (figures of 100 years or more are mentioned, though with some degradation of performance) and simple maintenance; the possibility of using direct or diffused light; and allowing modular construction, thus lending itself to all sorts of power levels, from the smallest to the biggest capacity. The disadvantages are the absence of a solution to the problem of large-scale storage of electricity,[8] the long way still to go before it is possible on a commercial scale, and the *cost*. Today the cost is about $20,000/kW(e) peak (sun at its zenith), or about $4/kW.h. From cells currently available a gain factor of 4 is possible, with a factor of 10 or more from other types of cells. The United States Energy Research and Development Administration (ERDA), now Department of Energy (DOE), whose optimistic forecasts are no longer valid (but it is such "errors" that get programs started!), had fixed successive goals, starting with $30,000 kW(e) peak in 1976, dropping to $2000 in 1985, and $100 in 2000 (in 1975 dollars). If these goals are achieved, ERDA estimated the electric capacity in the United States through solar cells would be of the order of 20 MW(e) in 1985 and 50,000 MW(e) in 2000! To put these figures in perspective, remember that the *world* market for solar cells was 0.75 MW(e) in 1977, the United States administration and army being by far the biggest customers.

The majority of photovoltaic systems currently use monocrystalline silicon (doped with boron and phosphorus to make a PN semi-conductor junction) with an efficiency of the order of 10 percent (or about 100W/cm^2 with bright sunshine). As the manufacture of such silicon cells is delicate and tricky, other materials are also being investigated, such as calcium telluride, cadmium sulphide/copper sulphide, or gallium/aluminium/arsenic, in polycrystalline or amorphous form, deposited in thin layers. Another solution, to diminish the investment involved in cells, is to reduce their number by partly concentrating the solar radiation (as with the thermal electric method). An inconvenient aspect is that concentration is not possible with diffused radiation, and the rate of activation by sunlight is not much above 30 percent (because of the shadows the concentrators cast over each other during the

[8] A possible storage solution was suggested at the International Conference on Hydrogen in Zürich, August 1978, by the coupling: photovoltaic electricity → electrolytic hydrogen, which could be stored in metallic hydrides. Encouraging progress has recently been achieved in the technology.

course of the day).

Concerning the cost of manufacture of solar cells it is sometimes said hopefully that it would only be necessary to make the same progress as was made with transistors for them to cross the threshold where they would become competitive. It might be added that large-scale production of photo-cells could be akin to the production of very complex structures with thin superimposed layers: colored photographic films.

Once these methods of manufacture have been developed, the other problems will perhaps appear simple, but nevertheless a solution will have to be found to them: reliability of components, economic storage of energy (which will possibly be the bottleneck);and reliable and economic converters to transform the dc current produced to ac.

Finally, it may be added that the oil industry is particularly interested in photovoltaic development; interested companies include Exxon, Mobil, Texaco, Shell, CFP and Elf.

The Solar Satellite Project. In every class there is an oddball, more or less sympathetic, more or less bizarre, who sometimes does very well for himself (at any rate, if one believes Mark Twain). Just as nuclear power has the super-nuclear island project of Marchetti and hydraulic power has the project of damming and evaporating a part of the Mediterranean, so has solar power its Glaser project. One could mention others, judged bizarre a few decades ago, that now are part of our everyday life. As not all these projects can be realized, they oblige us to some reflection, which is not a bad thing.

The Glaser project, constantly supported and financed by NASA, proposes nothing less than to look for solar energy where it is to be found in abundance, if not in the shopping centre of the sun itself, then at least outside the earth's atmosphere (flux of 1.3 kW.h/m^2), where it escapes from nebular variation and even, if judiciously sited some 36,000 km from the earth, escaping the alternation of day and night. It would be a power-producing geostationary satellite of 5-10,000 MW(e). Such a power satellite, weighing 10,000 tons, with 50 km^2 of photocell area spread out in space, would use ''amplitrons'' to transform the dc current produced by the solar batteries into microwaves of 2.45 GHz which would be received on earth by an antenna 7.5 km in diameter. Of course, some promising figures have been put forward, but hedged with so many ''if's'' that they do not justify too much time spent on them at present. As an invitation to reflect on the limits — some would say the hubris — of our technological civilization, this project is perhaps a gauge of solar energy's future, if not, on the other hand, the best example of what we should *no longer* try to do.

Until it was displaced by the steam engine, wind power was the motive force for many agricultural and industrial machines in England, Germany, Denmark, Holland, and France, not to mention sailing ships. This technique is still used for the production of electricity or the pumping of water by the isolated farmers of

Australia or South Africa. Current efforts are more ambitious and are concerned with high-capacity systems capable of being incorporated into existing electrical grids.

Various experts have estimated that the total energy concentrated in the circulation of the atmosphere is several hundred terawatts (a figure to lighten the hearts of the energy experts), but this figure does not mean much. What really counts is the amount recoverable from the layer from ground level up to 30 or 100 m, on the best adapted sites; and this fraction, which is unknown, like any self-respecting energy resource, probably amounts only to about 1 percent. This still represents a not negligible amount.

At a given spot the power that can be recovered varies proportionately to the cube of the speed of the wind. The efficiency of conversion to electricity is quite good, between 20 and 40 percent. As an example a generator situated in an area with an average wind speed of 7 m/s will intercept power of about 200 W/m^2, of which 40-80 W/m^2 would be convertible into electricity (rather like solar thermal electric conversion, but in principle taking up more land, although perhaps not exclusively). Wind power is particularly localized near the coasts of the temperate zone; it is stronger in winter than in summer, in the daytime than at night, and close to the ground rather than at altitude. It is regular when measured by the time scale of a year, a season, or an hour but very irregular by the minute or the day. The Atlantic face of Europe is well "ventilated," but the available energy differs widely between neighboring sites. Also, the windmill itself disturbs the airflow, and the major turbulence it causes is a disturbing factor on sites where one would like to use a number of units side by side or in cascade.

There is wide range of low-power machines used for pumping water or for generating electricity (1-10 kW), satisfactory for relatively isolated spots, but rather expensive, for example, about $1500-$20,000 for a 4 kW machine).

After the war Electricité de France experimented with two 1 MW machines, but work was not pursued following the arrival of nuclear power, which seemed more promising at the time (as regards scale and cost).

In the United States most of the development work has been concentrated on machines with two blades and a horizontal axis: an experimental machine of 38 m diameter in Ohio, with two identical additional machines of 200 kW(e) recently installed; two machines of 61 m diameter and 1500 kW(e) in the course of installation; and one machine of 91 m diameter on the drawing board. In Canada there is a machine of 37 m diameter and 200 kW(e) in construction, with a vertical axis, however.

For small machines, but also for big ones, the first problem is to reduce costs. For economic reasons, large machines will probably be limited to areas where the wind speed exceeds 5 metres per second. Among other problems one may cite those of protection from the risk of tempests, the irregularity of the electricity produced, presenting problems of storage (or mechanical or hydraulic power), accessibility of the site, and "acoustic pollution" (noise of the blades).

Nevertheless, wind power is still interesting when conditions seems particularly favorable (there are numerous studies on this in Denmark, for example).

Enormous quantities of water circulate in the atmosphere. They arise from the evaporation of the oceans under the influence of solar radiation; they are transported by the winds (another form of solar energy, as we have seen), fall back to the earth in the form of rain or snow, and finally return to the oceans via rivers and streams. So hydraulic energy — exploitation of the phase in which water from the atmopshere returns to the oceans, under the influence of the force of gravity — is thus a special form of indirect utilization of solar energy. It was also the first energy resource to be used after human and animal power. At the end of the nineteenth century it took on a nobler form and was rejuvenated as hydroelectricity. Increasing attention is being given to small installations, mini-power stations: in China, for example, over the last ten years 5,000 mini-power stations with an average capacity of 35 kW have been erected. In Europe, France plans to erect between now and 1985 about 4,000 mini-power stations (a little over 4 percent of the 90,000 sites recorded!)

It is also worthwhile to remember that the most powerful plants in the world are hydroelectric ones: Grand Coulee in the United States, with nearly 10,000 MW(e), Itaipu in Brazil, recently decided upon, with 12,500 MW(e).

About 23 percent of the world's electricity today comes from hydropower. The role of hydroelectric plants is in fact amplified by the fact that they contribute also to the management of water resources, allowing other non-exclusive uses, such as irrigation, municipal requirements, navigation, control of pollution, fishing, and leisure activities.

Hydroelectricity is a reliable and flexible source of energy, with the plant capable of being put into operation or stopped at very short notice. The plant has a long life and does not require much maintenance. As with other types of renewable energy, once they have been constructed, the plants are relatively insensitive to inflation.

Total world hydroelectric capacity in 1976 was 372,000 MW(e). The corresponding production of electricity being 1600 TW.h, it would have needed 325 million tons of fuel oil in thermal power stations to produce the same quantity of electricity. These 372,000 MW(e) represent about 16 percent of the ultimate capacity (installed and installable) estimated by the World Energy Conference in 1976. This percentage does not in fact include the small and mini-plants, which might add 10 to 20 percent to the global figure of 2.2 million MW(e), these 2.2 million MW(e) having a total productive capacity of 9,700 TW.h (representing 2 billion TOE used for thermal production).

This capacity is not equally spread between countries or regions. For instance the OECD countries taken together have already developed 46 percent of their capacity and most of their favorable sites, while other countries have not exceeded 7 percent. The world still has considerable potential for development. Even in the developed countries there is still scope for some expansion and for a better utilization of resources. Investment costs act as a brake on this development (these increasing as more and more difficult sites are developed), and there are sometimes some rather tricky conflicts with other forms of water utilization, with social, juridical, or political constraints such as international waterways.

However, it is probable that hydroelectricity will continue to develop strongly on a worldwide scale. It is forecast that in 2020 80 percent of the ultimate capacity may have been installed, with a productivity of the order of 7860 TW.h. Such a program implies enormous investment, more than 33 billion dollars (1976) on average every year, for forty-four years.

Tidal Energy. This is a form of water energy, but not of solar origin. Not justifying a chapter to itself, here seems to be the place to include it. Its world potential is relatively low (20-30,000 MW(e)) divided up between about 25 potential sites. One of the most interesting and important projects is in France, and could consist for example (there being numerous variants) in closing the bay of Mont-Saint-Michel with two dams, one near Cancale, the other at Granville, joining up at the Chausey Islands. This is a grandiose project, not only because of the potential power of 12,000 MW(e) (the equivalent of thirteen to fourteen nuclear reactors of the present LWR type) and a productive capacity of 25 TW.h, but also in terms of cost, estimated at about 30 billion francs (about $7 billion) and the scale of the work involved, a workforce of 10,000 persons being required for twelve to fifteen years. Nor would the ecological effects be inconsiderable. It is clear that the decision to embark on such an enterprise, periodically put forward, can be taken only by the public authorities.

Utilization of Biomass, or Biological Conversion of Solar Energy

Utilization of biomass is something of a ''return to earth,'' but a reconsidered return. It is the search for a rational employment of photosynthesis, attempting to direct it or even to imitate it. The possibilities and variants seem so numerous that the engineer (energy expert) can get lost among them, even more so because he is not generally familiar with such apparently simple techniques. A satellite is something calculable, but a harvest?

Photosynthesis is the conversion of carbon dioxide, water, and solar radiation into organic matter and oxygen. Organic matter consists of stored energy, stored work (with a match sufficing to release it into carbon dioxide, water, and heat!). The reaction of photosynthesis is marvellously complex, and its efficiency is very low, of the order of 0.5 to 3 percent.[9] The efficiency is low, *but there is storage;* with solar cells the efficiency is higher (10 to 15 percent) but without storage. It has been estimated that total production (energy content) by photosynthesis might reach 50 TW annually, six times current world energy consumption. Once more

[9] For example: 1.05 percent for beetroot in the Netherlands; 2.48 percent for sugar cane in the USA; 0.83 percent for maize in England etc.

the enormous size of this figure provides food for thought — all the more so because the products of photosynthesis can offer their energy in different forms and lend themselves particularly well to fuel production.

Sources of biomass are: forests (representing according to estimates between 2.8 and 4.5 billion hectares across the world, capable of producing annually between 8 and 12 billion TCE, taking 7 tons per hectare on average with 3,000 kcal/kg[10]), agricultural waste (5 billion tons annually, with 4,000 kcal/kg, or about 3 billion TCE), and molasses and domestic refuse (0.3 billion TCE annually?). Estimates of these resources are uncertain, and the possibility of their economic utilization on a large scale remains to be demonstrated. It is undeniable that they could play a not inconsiderable role in the energy budget, as in the developing countries where they are presently inefficiently utilized (for example, cattle dung is dried and burned with a very low efficiency instead of being fermented to produce natural gas *and* nitrogen fertilisers) or even simply looted (crises involving firewood and deforestation).

The possible ways of using wood are:

1. Combustion in air, going as far as incineration (wood fire)
2. Pyrolysis:

- carbonization in mills, yielding charcoal
- carbonization in a closed container
- total gasification, giving synthetic gas, which in turn, using Fischer synthesis, may be used to obtain alcohols, hydrocarbons, or ultimately hydrogen

3. Local production of electricity (3.5 kg of anhydrous wood, dried, would permit local production of the same quantity of energy as a litre of gas-oil derived from wood-gas)

As well as using "energy management" of forests (which already exists. but could be reorganized), another idea is the "energy plantation" based on cropping especially for energy purposes, such as sugar cane (for example, for the industrial production of methanol in Brazil as a substitute for petrol) or euphorbia. These plants secrete a milky juice known as latex, a complex hydrocarbon with a molecular weight of about a million; according to Melvin Calvin (winner of a Nobel Prize for chemistry for the discovery of the mechanisms of photosynthesis) it is hoped by grafting to bring the molecular weight down to about 50,000, corresponding to the molecular weight of certain petroleum fractions. One would thus obtain a sort of "gasoline tree" even more rewarding to cultivate than sugar cane, for unlike cane, this plant grows on infertile soil. This is evidently a crucial problem for energy plantations, which must not compete with traditional agricultural land, which is diminishing as the world population increases and as urbanization swallows up good land.

[10] 1 TCE = 7×10^6 kcal

The United States' Solar Budget

The table shows the United States Department of Energy's R & D budget for solar energy in fiscal 1979 and the budget proposed by President Carter's administration for fiscal 1980. There is an overall increase of 23 percent, but it is unevenly divided. The production of electricity via photocells will be effectively increased by 23 percent, while the biomass budget will go up 32 percent, admittedly starting from a lower base, and thermal-electric conversion by 19 percent. Although a prototype is under construction, the budget for exploitation of the thermal gradient of the oceans will go down by 11 percent.

To these figures should be added $169.5 million for 1979 and $155.6 million for 1980 for the Department of Energy's demonstration and commercialization plants, plus a further $159 million budgeted for 1980 by other federal agencies (such as Agriculture and the Tennessee Valley Authority,

Thus we arrive at a total budget of $714.8 million for 1979 and $844 million planned for 1980. Something to dream about.

U.S. R & D BUDGET FOR SOLAR TECHNOLOGY
(in $ million)

	fiscal 1979	fiscal 1980
electricity production using thermal means	102.1	121.0
electricity production using photocells	105.8	130.0
wind power	61.9	67.0
thermal gradient of the oceans	38.9	35.0
SERI (Solar Energy Research Institute, infrastructure)	3.0	27.0
biomass	43.2	57.0
program management and support	3.5	4.0
Total	358.4	441.0

Even after adding the possibilities of cultivating fast-growing vegetable and seaweeds, including blue-green seaweeds, which use an enzyme, hydrogenase, to produce photolysis of water, (its breakdown into hydrogen and oxygen), one has only touched on some of the possibilities of exploiting biomass. But before these can be realized — and certainly some of them will be, one day — there is still a long way to go. Systems of collection, conversion, and distribution must be studied, as well as the material balance and the energy balance, which may be very difficult. No more energy must be consumed by the various processes than can be recovered. Similarly, the material balance is concerned with accounting for all the materials and equipment used. Of course, the land balance is crucial, as the land requirements are inversely proportional to the efficiency. There is no doubt that these programs *must* be attacked in order better to understand them and to help arrive at logical choices.

SOLAR "SCENARIOS"

The preceding sections may appear very technical, but they seem indispensable to show the wealth and multiplicity of solar possibilities and variants, without glossing over their problems, and to indicate in what stage of development they are in. These thousand and one systems revolve around the sun just as many planets waiting to be inhabited.

But this multiplicity has two important consequences:

- It imposes the obligation of neglecting no possible path, with the current stage of knowledge. This is well illustrated by the French program, the second largest in the world after that of the United States, seeking to explore all paths. Although, however, its political orientation is unclear, for the program hesitates between giving priority to national needs — which it could not do, for this would be to disavow or deny the prior claims of nuclear power — and industrial deployment geared to the conquest of export markets. This is done with an eye to the developing countries, which is a good thing altogether, for solar technology is much less sophisticated than nuclear technology, for instance. In our opinion, it will be about a decade before some solutions emerge that are clearly more interesting than others.
- This multiplicity opens the way to all sorts of preferences. Not as in the early days of nuclear power, via a battle of competing systems, though naturally everyone defends his own with typically solar heat. In a new quarrel of the ancients and the moderns two principal trends seem to be opposed. On the one hand there is "soft" solar energy, very decentralized, à la Lovins,[11] having recourse to collectors, wood fuel, and domestic biomass with much reduced (voluntarily or compulsorily) energy consumption. Electricity, derived exclusively from water power, would be reduced to satisfying only some very specific needs, some 6-8 percent of final energy requirements. On the other hand there is "hard" solar energy, technological, à la Weingart, seeking to satisfy energy needs that have continued to grow, widely using secondary vectors such as electricity or hydrogen. This latter might possibly be produced by massive power stations implanted in very sunny desert regions (thus maintaining the interdependence of nations) or in ocean parks. It cannot even be excluded, though this is an extreme case, that recourse might be had to Glaser's satellites. This "hard solar energy" could be caricatured by saying that it cannot be

[11] Who has strongly influenced President Carter's new American energy policy.

distinguished from the "nuclear" world proposed a few years ago, just replacing the expression "fast breeder plutonium reactor" by "solar hydrogen superplant." This is an image that the "purists" of solar technology passionately reject. Solar truth, when it eventually emerges from the shadows, will doubtless be somewhere between the two extremes.

Altogether, this scenario poses the double problem of the speed of penetration of solar energy and of the possible levels of production. The question of speed of penetration can quickly be settled. Nowadays it is agreed that whatever path or paths are followed, time will be needed. Lovins speaks of fifty years for complete replacement of current fossil fuels by soft technologies. Others are generally looking at the horizon of 2050 (without wasting time!). Weingart speaks of seventy-five years being necessary before solar power covers about 50 percent of world energy requirements.[12] He notes, however, that this would imply a speed of penetration equivalent to or better than that of oil or natural gas in recent decades. This relatively slow process of maturing evidently poses the question of whether recourse to nuclear power during the transitional period will be necessary, or whether oil and/or coal resources are adequate to carry us over this period. We shall come back to this point.

Let us take heating. Admitting that these needs represent, say, 30 percent of a country's energy requirements, about half relate to densely populated urban zones, or where there would be insufficient surface area available for an adequate collector capacity. Because of logistic constraints, only half the market remaining could be catered for by solar heating, which, as we have seen, might cover 50-70 percent of heating requirements. (The rest would be provided by a back-up system.) In multiplying these percentages one obtains $0.3 \times 0.5 \times 0.5 \times (0.5$ to $0.7) = 3.75$ to 5.25 percent of the total energy balance. However, better ideas concerning construction along the lines of solar architecture might reduce heating requirements. Indeed, one could argue about any of these percentages, using reasons based on "market forces" and increase them, relying on mandatory measures.

Independent of logistical constraints (structure of the building industry, slow replacement rate for already constructed buildings, multiplicity of construction codes), there are financial constraints. Currently the equivalent price of fuel oil might reach, according to official calculations, 50 to 100 dollars per barrel before integral solar heating could become really competitive. These figures seem very pessimistic to us as regards solar heating. The problem is, as we know, quite different for hot water supply, where the premium cost is amortized in three to six years.

It is difficult to make forecasts for electricity, but we must note the importance of the host structure, the diversity of approaches, and the potential for improvements

[12] Needs having quadrupled or quintupled in comparison with current needs, reaching 30-40 billion TCE annually.

or processing (such as solar cells). The question is in fact closely linked to that of nuclear power, as we shall subsequently see.

As regards biomass, by extrapolation once more of market forces, possibly modified a little in favor of solar energy but not turned upside down or exaggerated, it is in fact very difficult, if not impossible, to make at present any global forecast estimate.

It is in this modest official context that we must finally place the all-solar schemes,[13] which are ambitious and which in one sense irritate us, for they oblige us to ask ourselves if they are possible, after all.

Few all-solar schemes have really been proposed. Because of this, it is interesting to consider as an example an all-solar scheme for France put forward recently, a very well constructed plan by the "Bellevue Group" (bringing together scientists from various disciplines from the Centre National de la Recherche Scientifique du Collège de France, Electricité de France, the Institut National de la Recherche Agronomique, and others, all professionally involved in research into renewable sources of energy). The Alter scenario assumes a low energy demand, compared with the majority of other long-term forecasts, and analyses the means of satisfying it with a combination consisting exclusively of solar plants. Some people consider the group is being completely unrealistic or at least idealistic, but it is maybe less so than they think. Some Electricité de France experts have challenged it by putting up a "science-fiction" scenario. SUN, standing for Scenario Ultra-Nuclear.

The final energy demand forecast for France in 2050, amounting to 141.5 TOE as against 146.5 TOE in 1975, would be that of a post-industrial society, an ecologically "responsible" one, having substituted for the categorical imperative of industrial growth that has ruled over the economy for two centuries an imperative of stabilization of productive capacity (and of the population, at 60 million inhabitants as against 51 million presently) with a standard of living that would nevertheless be comfortable for everybody. It might ultimately be much more difficult to get such a balance of energy *demand* accepted — and, as the Bellevue group is well aware, it is hard to see what political group could impose it — than it would be to put the scheme of solar supply into operation. Whatever it looks like, and despite the total balance of consumption being similar, the division into sectors will be appreciably different from that of today. The residential and tertiary sectors will be more important, and those of transport and industry will be reduced. Apart from heating, the scheme involves the production of solid agricultural or forest fuels in convenient form (such as granulates) or fuels derived from biomass, with a relatively significant amount of electricity production (contrary to Lovins-type scenarios) with various units producing electricity directly, thermal electric or photovoltaic plants, supplemented by water- or tidal-powered units (a *small* Mont-Saint-Michel project) or wind-powered units.

[13] Schemes, not scenarios, for they do not go as far as describing the various stages before reaching the final objective.

- solar heating would supply 80 percent (34 M TOE) of commercial and residential heating needs, and 40 percent (11.5 M TOE) of industrial heating needs, using 250,000 hectares for facilities which would be widely scattered.
- solar power stations producing electricity directly (13 M TOE) and producing hydrogen by electrolysis (14 M TOE) would take up 450,000 hectares.
- biomass would be transformed in units of medium size (150,000 TOE/a) into solid fuels, supplementing domestic heating and particularly industrial heating, and into liquid or gaseous fuels other than hydrogen. The plants producing these fuels would be located near agriculturally productive or forest sites, one plant supplying several countries. This biomass utilization would entail the permanent exploitation of 5 million hectares of forest (compared with 15 in 1975) and 2.5 million hectares of agricultural land (compared with 35 in 1975).

These quantities are certainly significant, but not prohibitive. They represent a little less than 6 percent of the national territory — the road network alone uses nearly 1 percent — and only a fraction (less than 10 percent for sure) would be totally industrialized.

What if this "scenario" were possible after all? The scientific attitude would be not to commit itself today, but it would no longer be wise not to study this scenario. According to other scenarios we shall have to use an increasing proportion of our income to satisfy our energy needs. According to this scenario or other similar solar ones,[14] a not inconsiderable portion of our patrimony is assigned to our energy needs. It is a question of choice.

[14] We ourselves in the course of various studies at IIASA have explored an all-solar scenario, in its choice of units quite similar to that of the Bellevue group, but assuming nearly double the energy demand and perhaps with a little more optimism as regards the potential of technological progress.

6

THE NUCLEAR APPLE AND THE SOLAR ORANGE

The energy resource par excellence is oil. Concentrated just as much as necessary, easy to handle, suitable for storage, easy to put into operation and use. But, unfortunately, difficult to find, more and more, ultimately exhaustible, with the cream hastily being taken off the deposits. Then comes coal. Also concentrated just as much as necessary, but harder to handle, less easily stored but flexible in use; its principle trump is the size of its reserves and resources.

Water, so vital for our civilization, is not lacking in energy potential. Cold, coming from the mountains, it drives turbines; hot, coming from the earth, it heats homes. However, these capacities do not cover the needs we have created.

With our current state of knowledge, there seem to be two possible ways offered for us to follow, as regards energy in the future:

- nuclear energy. In the widest sense, currently meaning primarily nuclear fission and, by "annexation," thermonuclear fusion. Highly concentrated (especially with breeders), so concentrated even that problems of handling upstream of the reactor hardly arise. Easily stored, because of this same concentration; but somewhat lacking in charm because of the reactor. This requires a sophisticated food that involves a long chain of complex and expensive preparations. This food is ill digested, and what is rejected presents formidable problems of handling and storage, such as have never been known in the past.[1] This nuclear operation, the apex of a slow process of increasingly complex technology produced by man and society, gives rise to anxiety, as does anything extreme.
- solar energy. With the sympathetic face of a forgotten god, bearing no ill-will, but not in a hurry. Contrasting with oil, because of the enormous resources, and at the same time comparable to it in the multiplicity of its applications. Contrasting with nuclear power (highly concentrated) by its diffuse and seasonal character, but resembling it by virtue of its perennial character. Its

[1] Formerly, man stored only food and treasure.

"benevolent" rays seem to stand against the threatening rays of the atom. Super-super-coal by its abundance, this so simple form of energy is, alas, not so simple to put to work. So what, one might say; nothing is free in this wicked world; which in terms of the language of energy might be put as: there is apparently no miracle energy.

How do we balance the nuclear apple against the solar orange? For some people it is a matter of cost, for others a matter of taste. In fact, there is no single simple answer to this problem.

APPLES AND ORANGES

As a general rule it must be said that you cannot balance apples against oranges, or nuclear against solar energy. The simplest method would be to compare the price of identical finished products, for example calories (because they depend on temperature) or kW.h. But it is difficult to establish common and compatible cases; the same items sometimes appear in different forms, or do not appear systematically, or are accounted for in a different manner, for example, as regards their possible amortization. Moreover, it is already difficult to compare costs and, *a fortiori,* prices, including the various taxes (which can reflect different fiscal intentions) for such a topical product as a kW.h. It is even more difficult to compare costs or prices as they may be at the end of the century, in 2025 or 2050[2]. It is not always exactly known today what is the real cost of a nuclear kW.h, because certain downstream parts of the fuel cycle (reprocessing and waste management) have not yet been fully defined and have not yet become really commercial.

How, then, can we compare what the prices or costs may be in twenty or fifty years with the possible price of a solar kW.h produced by solar cells whose nature is as yet not definitely known and which are manufactured by processes that do not yet exist?

Here we are evidently touching on one of the essential difficulties, not the only one, alas, of all studies on the future: the impossibility of making valid forecasts of the evolution of techniques *and their real costs*. Nuclear power, the newest of all the new forms of energy, teems with past, present and future uncertainties; they can also be seen in more extreme form in the thermonuclear equivalent. Just as many can be expected for solar power at present and in the future.

Coming back to cost comparisons, let us recall that the United States system of light-water reactors was "preferred" in Europe to the British and French gas-graphite system because it "seemed" to offer an economic advantage of the order of 10 percent. In reality, and happily, this was not the only argument, but it

[2] This is a well-known, insoluble problem: that of determining discount rates over a long period, and the no less delicate problem of rates of inflation.

was this that we were led to believe was the most decisive one. The future was to reveal subsequent mushrooming of costs, exceeding 100 percent, and quite unanticipated. This shows the fragility of such comparisons, sometimes abused by economists (including, among others, some American economists faithful to the standard dollar). One should not go to the other extreme and exclude such economic comparisons, in the French manner, as is sometimes said (cf. the Concorde story).

Instead of comparing the costs of products, one can widen the comparison to the total cost, even the national cost of the system as a whole, of the envisaged alternative source of energy, in terms of direct and indirect investment (for example siderurgic-steel factories to produce the steel necessary for the fabrication of nuclear reactor pressure vessels), and similarly try to calculate what the estimated benefits are. This is the principle of the widely used "cost-benefit" method. It is somewhat like trying to replace the sum of micro-errors with one macro-error.

Where things come to a head is when one tries to include social costs (a problem brought into the limelight by antipollution policies) or advantages difficult to express in figures, such as the beauty of a landscape or view or the appreciation of silence. The method is derived from economics and is applied to a noneconomic field, leading to long lists of studies with hardly quantifiable impacts (migration paths of the caribou with relation to the Alaska oil pipeline) and, of course, hardly comparable. What relative weight should be given to two such different attributes as the purity of the air and the survival of already rare aquatic species, which however form a link, possibly a fundamental one, in the ecological chain? Again, it is a question of apples and oranges.

These lists and studies of environmental impact in the widest sense are difficult to use conclusively. On the other hand, in their optimal form, they are offered to the decision-maker, a person much could be written about. But this, alas, would take us much too far. The decision-maker will be swayed by his preferences, consciously or not, and will apply his "objective function," consciously or not. This objective function could be to minimize the actual financial costs (the commonest criterion!), or to minimize pollution, or to optimize the management of natural resources, or maximize employment. Finally, the cost-benefit of impact–matrix expert will perhaps approach the decision-maker, or the latter will take further steps to attain a better appreciation of the problems by trying to establish "preferential functions," emerging from preferences (and compromises) between various attributes, expressed preferences of identified groups such as consumers, producers, legislators, and environmentalists. This complex alchemy of preferences is one of the latest tools of "multi-criterion analysis," the scholarly name applied to balancing apples and oranges.

After these scientifico-economic or pseudo scientifico-economic methods, one may well ask if the final and democratic stage of such a comparison is not ultimately a popular vote, a referendum. It is a big subject, true, both in itself and in its possible application to the choices and alternatives as regards energy. Such a solution was proposed in various countries, and Austria had recourse to it, indeed.

In this book we shall limit our contribution to the "balancing" of the nuclear apple and the solar orange to establishing a nonexhaustive list of attributes that seem important with regard to energy options, nuclear or solar. From what has gone before it will be understood that the order in which these attributes are listed does not necessarily reflect a scale of values. If it did, it would necessarily be subjective, like any comments offered for the reader to ponder on, aiming at possibly questioning certain received ideas. Everyone can modify this game, or rather this grid, by adding to it or changing it in the light of one's own knowledge, convictions, perhaps mood, or even caprices. One may possibly arrive at conclusions diametrically opposed to those sketched out here.

THE STATUS OF DEVELOPMENT OR MATURITY

Nuclear power, as has been seen, is more advanced than solar power, with a lead of twenty years at least. This is of course due to the enormous development effort from which nuclear power has benefited during and after the war and for warlike purposes. It is available today for the production of large quantities of electricity, corresponding to the scale of the needs (even if the ultimate legitimacy of these may be questioned) and technically, say, on the gigawatt level per unit plant. It could possibly become available tomorrow for thermal uses, although somewhat handicapped by the too-large scale of its units. It would technically be easy to reduce this, but probably in exchange for a big degradation of its economic advantages. Current light-water reactors and the fuel cycle upstream are utilizable as they are. The fuel cycle downstream is less advanced, but there is no reason (as France is seeking to demonstrate) why the technical problems should not be solved.[3] On the other hand, reactors other than the light-water type still need much effort to be brought to their full development, though curiously enough it is the most difficult types (sodium-cooled fast breeders) that are most advanced within the wide range of alternatives.

Solar power is available today for certain thermal uses, but its total contribution to the energy balance will remain modest for a long time. Experiments in progress (the Albuquerque project and others) should permit decisive progress on the road to solar electricity from the next decade onwards, and there might be a breakthrough in solar cells at any time. Contrary to the situation with nuclear power where it is uncertain whether the current size of reactors represents their ultimate size (with talk of 500 MW(e) reactors in the year 2000 or 2020), solar plants presently planned may already be close to the optimum size of module, without having to go through the various stages from megawatt to gigawatt (a factor of 1000), as nuclear power did.

[3] This is not to say that the operation will be profitable, or without risk. The question as to whether the risks are acceptable and/or accepted is dealt with in another section.

EFFECTS OF SCALE AND SERIES

As unit size increases, economies of scale are sought, among other things, and these are harder and harder to achieve as size increases.[4] Also, at a certain moment, which nuclear power maybe seems to have reached, it is necessary to weigh carefully economies of scale in terms of production with "diseconomies" of scale in distribution. One of the reasons why it is hard to benefit from economies of scale is that a very small number of power plants are designed and built more or less one at a time or in a very small series (as regards major components). Attempts by the customer or the manufacturer to standardize are not easy to realize. They run up against the principles of competition, and tend to impede technical progress, which mostly proceeds an inch at a time rather than in jumps or revolutions.

In a sense, effects of scale can be opposed by effects of series: the producing unit is large-scale, but the scale of the object produced in large numbers is relatively small. One example, perhaps the best one, is the automobile industry: it is not always realized that the current cost of producing a car is of the order of $2000 ($2 to 3 per kg), not including of course taxes, financing, bonuses, and so on. This is an exceptionally good example of economic efficiency and technical organization. Such an example must, or ought to, inspire solar energy now or in the future.[5] And why not?

COSTS, PRICES AND INVESTMENTS

We have already quoted some costs and prices, and we have also said what we think about these. Having made this point, we may note that the move from conventional oil to nonconventional oils, from raw coal to gasified or liquefied coal, and from classical fossil fuels to new forms of energy is being accompanied by a parallel move towards higher and higher investment. It is this investment, greater than that associated with current practice, and still a long way from equilibrium values, which we must seek to compare.

As we have seen, such a comparison seems at the moment favorable to nuclear power (as regarding in any case investment costs, which form a decisive part of total costs in the capitalist economies of today with their high interest rates).

Independent of their level, it is also important to give equal consideration to the nature of the investments and the investors. In the case of nuclear electricity it is mainly public or private utilities who make the investment, often benefiting from favorable financing terms, though less so today than formerly, over a long term. These conditions could perhaps be extended to a very centralized solar plant, as on a municipal scale. It will not apply to individual solar installations, unless new and extensive financing methods can be found from the outset.[6] Vague suggestions

[4] Some "worrisome" studies by J.C. Fisher (of General Electric) cast considerable doubts on these scale effects.

[5] One could contrast the nuclear type of passenger liner with the solar-driven car.

[6] We were surprised, nevertheless, by the number of cases in which potential users were prepared to pay a large initial investment premium for solar energy.

have been made about the banks playing a part (as is the case on rare occasions in the United States) or utilities producing or distributing electricity or energy (that is, on a leasing basis).

The user, struck more obviously by the price per kW.h or litre of fuel, does not always realize the size of investment imposed on the energy producers. To light a 50 watt bulb (costing 50 cts) by simply pressing a switch, over $50 must have been invested in production (a hundred times more) and just as much or more in transport and distribution. It is this $100 *per bulb*, with the "premium" solar cost possibly being compensated for by the absence of distribution costs, that the producer of solar electricity must be prepared to invest.

Finally, in connection with costs and prices, it is also worth mentioning the tax problem. Currently the level of tax on heating or industrial fuels (including uranium) is relatively low compared with that on motor fuel, where it is heavy (contributing about 10 percent to the national budget of many European countries). When the production of future motor-fuels, hydrogen from electronuclear electrolysis, or methanol derived from biomass becomes much more expensive, the whole tax question will doubtless have to be reviewed, but for the time being it is not known in what direction. Similarly in the case of a changeover from a highly centralized energy system to a highly decentralized one.

Thus, considerations of cost and price, though important, cannot be decisive in a long-term comparison between the nuclear apple and the solar orange, which we knew anyway.

CENTRALIZATION OR DECENTRALIZATION

The decentralization of solar energy is its proudly demonstrated merit, and it is also part of its stock image, partly true and partly rather oversimplified. Is a totally decentralized society as regards energy possible? It seems not. Is it desirable? Nobody knows.

It is hard to imagine an industrial or post-industrial society just consisting of self-producers who are also self-consumers. How far should decentralization go, and how far centralization? The latter is perhaps worse in practice than in theory.

Nuclear power, the precursor of an industry placing more and more reliance on complex technology, is also accused of being excessively centralized (lending to the rich . . .). But the production of steel is centralized, so is automobile production, not to mention many other examples. The question for nuclear power is really whether it is the cause or the consequence of excessive centralization; whether it is the motor or the symbol; whether it must be destroyed because it is the motor or pitied because it is the symbol (pitied, because it can never be examined dispassionately).

It is perhaps to combat this accusation of excessive centralization, the consequence of the tendency toward gigantism of individual units and the network connecting them, that some nuclear scientists have investigated the prospect of a certain decentralization of nuclear power by the proliferation of small-scale units. We think this is rather like trying to pass a camel through the eye of a needle. In practice, nuclear power calls for large-scale and concentrated installations, producing massive amounts of energy and associated products in giant energy parks. The question often put forward nowadays concerns the "democratic" control of such centralization. In as far as this question is sensible it must be put not just as regards nuclear power, but as regards the entire system of production. Again one comes up against considerations of society's choice.

As regards solar energy, the models proposed are still too vague for it to be possible to appreciate the strengths and weaknesses of its decentralization, the limits of this, and its integration into industrial or post-industrial society as a whole.

ACCEPTANCE, RISKS, PROLIFERATION

As regards these, the scale seems to be tilting away from nuclear power, from whichever side it is approached.

One may ask oneself today whether public opposition to nuclear power will continue to increase, level out, or even decline. One may also ask if it is well-founded, or what is its motivation, sometimes. But it cannot be denied or its importance underestimated any more than proponents of nuclear power can claim to have a monopoly on reason.

Our civilization has become crochety and argumentative. With nuclear power the argument has become national, even international. With solar power, if it fails to escape, the argument will perhaps be on a municipal level, the village pump level.

For the time being, solar power is undeniably benefiting from a favorable attitude, so favorable that, as we have said, some people are prepared to pay a "premium" to profit from it, partly frustrating the economist's forecasts. Will this last for long? We may expect it will. But some of us may remember, not all without nostalgia, that nuclear power also benefited initially from a certain infatuation.

Nuclear power is potentially dangerous. But it is not always easy to *measure* or to appreciate if the dangers of plutonium are so much greater than those of phosgene, of carcinogenic substances produced by the ton in the gasification or liquefaction of coal or lead (or mercury, or other metals) and serenely, irresponsibly vented into the atmosphere or the ecosystem — not to mention the "explosive" character of a world which would annually handle several billions tons of liquefied natural gas. The existence of risks does not mean they must not be run, and necessity may impose sacrifices. Such sacrifices are in vain and to be condemned if the same advantages can be achieved using other means, such as solar energy. Here there is a ray of hope, in any case, and this is what it is vital to try and demonstrate.

As regards the proliferation of nuclear weapons, inevitable in the absolute sense, it is of course facilitated, we would even say encouraged, by the proliferation of civil nuclear programs. Nuclear technology is unfortunately more closely associated with war than coal and steel have ever been.

The "dangers" of solar energy are currently little more than anecdotal: a concentrated solar bundle getting off target (though remaining concentrated!), hitting something nearby and melting it on the spot; or, though properly aimed, being intercepted by an aircraft, burning its wings in the bundle (but what was it doing there in the first place?). Less well known, but more serious, are the climatic risks on a global scale such as variations of the albedo[7] over large or localized areas. Local climatic changes are not typical side-effects of solar exploitation, but are more or less linked to any production and/or geographical transfer of energy over a certain level, no matter whether its origin is solar, nuclear, or fossil fuel.

Is there such a thing as solar "waste"? In a sense, yes. It is the waste associated with the major mining operations necessary to obtain the enormous volumes of material required in the manufacture of solar plant (especially heliostats). The nature of the problem is not as serious as the irritating problem of nuclear wastes, a ball and chain for which today there is no key. The necessity for management and control of nuclear wastes over centuries, over millennia, has been compared to the need the Dutch have always to watch their dykes, in order to avoid a national catastrophe. This watching commitment has also been compared to mankind's obligation, unconscious and collective, always to cultivate agricultural land. If the carts were to stop, if the peasants withheld their labor, then mankind would be decimated by hundreds of millions, quickly reduced not to the return to earth of which the neo-naturalists sing but to gleaning and hunting. The comparison is not valid, as the methods of surveillance of nuclear "cemeteries" require, given the unfortunately insufficient amount of knowledge we possess, an engagement, even a bet, on the lasting character of our institutions[8] that one may and should normally be hesitant to make.

This shadow, this threat, as some call it, justifies and demands the greatest prudence. For the same amount of services rendered, solar power, if it could meet our needs, should be chosen unconditionally unless there is significant progress in the management of nuclear wastes, that is to say, progress sufficient to allay the majority of anxieties.

Lastly, we must also mention (without elaborating) the question of nuclear terrorism. Like everything to do with nuclear power, here too it is easy to come to an extreme, an end point. There are certainly all types of targets for terrorists: chemical factories, gasometers, big dams, or even a simple football stadium on the occasion of a world cup final, again fruits of concentration and centralization and a certain kind of industrialization. Nuclear terrorism, as it has occasionally been described, exceeds in horror all the others, as does probably the panic it would cause or the blackmail it would permit. Could our society protect itself against this

[7] That is, of the earth's capacity to reflect solar radiation, a key factor in climatic equilibrium.

[8] Compared to this requirement the lasting character of the Catholic Church is just a pale example, with its "mere" twenty centuries of survival and evolution.

threat without becoming something of a police state? Have the risks of nuclear terrorism been exaggerated? It is difficult to reply. What can be incontrovertibly ascertained is the escalation of violence all over the world. Just as society, thanks to the suicidal lack of thought of its richest members, tends to engender in towns a minority of protesters who will stop at nothing and on a world scale, generations of "Palestinians" or the "desperados" of civilization.

Less tragic, perhaps, but nevertheless threatening, are the demands for nuclear "quality" and "surveillance" in a world where laissez-faire attitudes and irresponsibility daily gain ground. In one sense, and this is personally our greatest fear, society seems to be lining up against the strictness, seriousness and order needed by, *demanded* by, a nuclear society.

SUPPLY AND NATIONAL INDEPENDENCE

From the technical point of view, solar and nuclear power can be largely national undertakings, users and multipliers of national grey matter.

As regards nuclear power, the breeder reactors burning granite from the mountains or, more precisely, the few ppm or parts per million of uranium that it contains, would reduce the problem of supply, not in terms of volume (which would be enormous) but in terms of political independence. This solution, or rather this picture, is theoretical and distant. The uranium needs of breeder reactors are undeniably small, but it would be necessary to see they were assured. While waiting for breeder reactors and limiting ourselves to water reactors only for a scale of fifty years or more, we are not very sure from what sources our uranium needs will be met. European countries, like many others, will have to import a significant amount of their requirements, increasing as their programs of electronucleariza-tion become more ambitious.

The countries producing and exporting uranium today are: Canada (with a strict policy), Australia (with changing policy), South Africa (threatened in the future), and the African states (in the middle of an Africa in the course of change, the object of greedy desires and intervention from outside). It is perhaps a pessimistic picture, but the subject of concern and even simple question marks are not lacking. It is less the possibility of an uranium cartel that should be feared (which already seems to exist, though in a much more veiled and hypocritical form than OPEC, the oil producers' cartel) than interruptions in supply as the result of violently developing situations.

As regards solar energy, independence of fuel supplies is evidently total. This is not necessarily the case, however, as regards the materials, such as copper or cadmium,[9] for which it is hardly possible today to estimate future needs, given the

[9] Cadmium is used for a certain type of solar cell; it is evident that the decision as to what type of solar cell to put into commercial production must take into account the world availability of the constituent materials. Copper is the noble metal of electricity. If necessary aluminium could be substituted, as it is much more abundant. The choice of solar materials is an important problem, because these are used in large quantities.

immature state of development and domestication of solar energy.

In the case of various solar scenarios, one may also imagine situations where countries with insufficient sun, like many of the European countries, have resources to the import of fuels, hydrogen or methanol, made in countries with an excess of sun like, for example, numerous African countries. The solar installations producing these fuels might be owned by the consuming and importing country (concessions and sun "dues") or to the country producing and exporting, or to both, as in some cases involving the transport of liquefied natural petroleum gas.

As regards energy dependence, two solutions are possible to limit or suppress it: independence, a big word and a big "carrot," and interdependence, recognition of the solidarity of nations. One may ask whether the former, much touted in certain quarters, is the more desirable, without raising the question of knowing whether it is possible, or if it is sensible, limited only to the supply of energy. The same question here posed with regard to solar power is relevant as regards nuclear power.

ENERGY AND NATURAL RESOURCES

There is too much of a tendency to consider energy in isolation, outside the social context on the one hand, and outside the natural environment on the other. In fact, when one consumes a ton of oil, one is also consuming the other energy invested or utilized in its production, transport, refining, and distribution: the water used or consumed in these processes, some land, immobilized or occupied by factories, or temporarily destroyed by mining operations. One is consuming other materials — steel, iron, copper — constituents of the equipment used. All this is operated by manpower, often very diversified. In other words, one does not consume just oil, or energy, but a complex of natural and human resources.[10]

The consumption of "WELMMITE" tends to increase as one becomes oriented to the use of more diverse forms of energy resources, and as energy consumption itself increases. It is thus necessary to watch out that the energy problem is not solved by creating scarcity elsewhere along the chain of natural resources.

Everything depends on the economic value put on these resources. A little reflection will show that resources, whether utilized by man or created by God, only become resources in as far as they are used by man, and hence in as far as some economic value is attributed to them. Before the discovery of radioactivity and the development of atomic energy, the mineral uranium was practically a valueless rock.

These natural resources used for energy or other purposes are naturally of very

[10] Called by the author WELMMITE (Water, Energy, Land, Materials, Manpower), the subjects of his WELMM method, developed at the International Institute for Applied Systems Analysis in Laxenburg, Austria.

different kinds. However, they interact closely, and these interactions must not be forgotten, whether one is concentrating on energy or not. Natural resources are also threatened in various ways by shortages and/or overexploitation. Do we not regularly hear about the water crisis, the land crisis, and the shortage of this or that material?

Water

On a world scale, at every moment there corresponds to each human being "on average" (water, unlike energy, not usually being consumed, but used and not destroyed) some 7,500 m^3 of fresh water (excluding the oceans). Nevertheless the availability of water is a growing problem in numerous conurbations or regions of the world (the Sahel zone, for instance), which could be aggravated by energy production (competing with other uses such as irrigation in New Mexico or Wyoming for the production of coal and uranium, their final conversion, and the restoration of disturbed land). As the potential of rivers has been exhausted, the enormous needs of nuclear power stations have led to their migration toward ocean shores, with the accompanying disadvantages and penalties in terms of transport. Thought is now being given to bringing them close to centres of consumption, even if this entails using other methods of cooling (more expensive by far) such as cooling towers.

Land

On a world scale there are about 37,000 m^2 of land per inhabitant "on average", of which only about 10 percent is arable land. If the entire world population were concentrated in the United States, the population density would be about the same as in the Netherlands. Nevertheless the land crisis threatens a number of industrialized countries, without mentioning the soil crisis (the biovegetal covering), soil being without doubt a more valuable resource to humanity than oil, like it nonrenewable, but also without any possible substitute.

At the moment the biggest threat to land is due not to energy but to urbanization, to industrialization, and to the development of networks of roads and motorways. To take an extreme case, it has been calculated that in California every further 1000 inhabitants "cost" 96 hectares of arable land, which are covered by concrete. If there is not a big change in this traditional trend, between now and 2020 half the arable land in California will be covered by concrete. Not to mention the concreting over of the French Riviera.

Energy "consumes" land through mining and plant. The former is transitory, given the current policy of land restoration and *soil* restoration, but during the few years of exploitation the effect is so devastating that it is becoming more and more difficult to open new mines. It is often said that in the United States it is not so much the mineral resources that are lacking, but the possibilities of getting at them, of exploiting them. For by its policy of increasing land reserves, the American government is progressively reducing the land open to exploration and exploitation by mining, a phenomenon reinforced by public opposition to despoiling the countryside. This phenomenon is tending to become more general all over the world. Apart from the mining problem, it is hardly any easier to find acceptable sites satisfying even a certain number of specific criteria, let along acceptable sites for big energy plants, including big electric power stations. It is an important problem that all utilities have been facing for several years in their long-term plans for rationalization and selection of sites. In this respect, nuclear power (especially breeder plants) are economical of land, with the mining implications being minimized because of the concentration of the fuel and the extensive use made of it.

Solar energy on the other hand, ''gobbles up'' land. So careful plans must be made, especially in possible agricultural energy applications, which might tend to compete or conflict with conventional agriculture.

There are various forms of opposition to nuclear sites. We do not know what forms of opposition to, or argument against, solar-electric sites might arise, big fields of mirrors pointing their towers at the heavens, or vast arrays of solar batteries.

Materials

The average cubic kilometre of the earth's crust contains about 200 million tons of aluminium, 130 million tons of iron, more than 200,000 tons of nickel, and nearly 200,000 tons of copper. Yet there are periodical fears of scarcity, as threatened in the first report of the Club of Rome. Without going as far as the average yields of the earth's crust, it may be thought that mineral resources are not limitless, but in general sufficient to satisfy all our various types of long-term needs, with some metals perhaps being less abundant or more critical than others, but not without possible substitutes(as with mercury, for example). Though one may be reassured about the size of the resources in the ground, fears are justified concerning accessibility of these resources (the problems of land just mentioned) and concerning our ability to produce and/or commercialize these minerals in time. These problems lie at the heart of the North-South dialogue, which, alas, does not yet seem to be leading to a happy and necessary solution.

As regards the recycling of materials, which seems to be becoming more important as time goes on, with the continued growth of the world's population, and the expected industrialization and development hoped for by the Third and Fourth Worlds, solar energy fits in much better and much more easily than nuclear energy. Not only are some nuclear materials difficult to recycle, but they also effectively contribute to aggravate the problems of wastes. They pose the difficult question of the future fate of the installations: should they be buried in concrete, made into pyramids at the rate of several a year from the beginning of the next century (for the reactors starting up today)? Or should the facilities be completely dismantled, which is technically feasible, but would impose a heavy economic burden on future generations?

Manpower

We shall limit ourselves succinctly to two aspects. The first is the possible risk of the lack of skilled manpower for operations such as mining (direct, for fuel, or indirect, for the materials necessary for various kinds of equipment), including when these operations leave "the surface" to return to "the depths." It seems that not sufficient account is being taken of this constraint of the third era of civilization: after the civilization of Homo sapiens, then Homo economicus, the civilization of the "intelligent mole," when man will spend part of his time digging holes and then filling them in again.

The second aspect is the dramatic problem of general unemployment arising all over the world endemically, growth or no growth. A prime objective of society should be to maximize employment. Energy policies should bear this problem in mind more than they currently do. This is not easy with our current state of knowledge of, for example, possible solar scenarios: the unknowns include the amount and quality of manpower necessary to build, operate, and maintain solar energy installations.

To conclude these reflections on the systems aspect of energy and natural and human resources, it may be said that not only must energy be integrated into the economic system, but it must be considered as just one of a number of different natural resources. These resources must not be considered in isolation, but must be the subject of overall planning and management, because of their permanent interactions.

PRIMARY, SECONDARY, TERTIARY ENERGY

For various steps of the energy chain between mine and user, even simply in order to utilize potential "fuels" that cannot directly be burnt (such as uranium) it is tempting to attempt conversion to a better or best adapted form. But with each conversion there is a loss of efficiency, and this amounts to buying optimization in terms of the commodity at each state at the expense of descending the scale of ultimate efficiency, which hits at the ethics of natural resource management and tends to encourage the spread of a civilization that loves comfort, ease, and even a little laziness. Electricity is the best-known example, and for this reason it has become a favorite target for Lovins and Co. But the development of new forms of energy, some not directly utilizable or only in certain sectors, is widening the debate to other vectors of secondary forms of energy, tertiary forms, quaternary forms, hydrogen, methanol, synthetic gas, and so on.

Natural gas is an almost ideal fuel. After a little processing during production it is delivered just as it is to the consumer. This alone suffices to explain its success.

The same applied for a long time to coal, though it was much less easy to handle. Given the abundance of this latter and the convenience of the former, the idea naturally arose of "synthetic natural gas," bought at the expense of some loss of efficiency.

Oil remains an exception. It can be burned just as it is. But most often, at the cost of a relatively simple refining process, it is "split" into a thousand *secondary* products that can nearly all be utilized directly: heating fuels, motor fuels, and the like.

This beautiful simplicity vanishes when it comes to nuclear power. The treatment is complex, and the usual product, electricity, currently meets only some 10 to 12 precent of our final needs. As we know, there are two paths that may be followed if nuclear power is to have a future: to increase the sectors using electricity — initially easy, subsequently more and more difficult, even problematic, as the electric car suggests — or to move from electricity to a tertiary vector such as hydrogen, with a further loss of efficiency. As regards the direct utilization of nuclear heat, the excessive size of reactors militates against this: the mountain gives birth to a mouse and crushes it. These faults will be even more aggravated with thermonuclear fusion insofar as the plants look like being even bigger.

Geothermal energy, as currently conceived, invites the use of secondary forms of energy if its range of uses is to be extended, but the potential forms of secondary energy (as ever, headed by electricity) will be produced with a very poor efficiency in view of the low temperatures involved.

As for solar energy, this is more like oil, with hundreds of ways to capture it and hundreds of ways to use it: direct low, medium and high-temperature heat, possibilities of producing electricity, possibilities of fuels, and all these in discrete amounts corresponding to most requirements. With solar energy the concept of energy yield loses something of its sense, as the source is practically infinite. It regains some of its sense when the yield from materials, land, and so on is considered, that is, the management of natural resources. A no less attractive aspect of solar power is the possibility of relatively short chains or possible interest in coupling the complementary forms of solar energy and oil more closely.

NORTHERN ENERGY AND SOUTHERN ENERGY

Unfortunately, nuclear energy, which is better adapted to the highly industrialized countries where it was born, adds to the rift with the southern countries, to which it has not been adapted despite all the campaigns (all too often for commercial reasons) seeking to do this. Either the developing countries buy "everything." as Iran did during the Shah's rule, which is naturally interesting to sellers ready to sell "everything," even armaments or "sensitive technology" as far as they are allowed to do so. Or the developing countries try to do everything themselves, like India, stumbling along and wasting energy that could certainly be better utilized in other sectors. Iran and India are large, relatively developed countries. What about the others, with a range of sentiments swinging between wounded pride to passionate nationalism, leading them to want to join the nuclear club, with military vanity not being totally absent?

Solar power would undeniably be better adapted to these countries. It can only be regretted that the egoistic North has not established a vast solar Marshall plan for the South, which would also be a way of rewarding it in the long term for the irreversible inroads made on its energy resources (oil and gas), whatever price was paid for these.

THE JUDGEMENT OF SOLOMON

Of all the elements in our grid, some are relatively clear-cut (the disadvantages of nuclear power with regard to risks, wastes, proliferation, disadvantages of solar energy with regard to the management of natural resources, where it is more greedy); others are less clear-cut (the economic advantages of nuclear energy, possibly declining in time). One could say that all these call for more detailed study and/or closer comparison.

All this leads us to say that it is urgent to wait. Without going so far as to stop all nuclear programs, we think any expansion of them should be avoided. *They grew big because of panic; the time for reflection has come.* As regards breeder reactors (fast, or possibly thermal) and where nuclear power would be effectively pursued long term, their advantages balance their disadvantages: for instance, it is desirable to finish the French Super Phénix without haste and to "play" with it, *before* deciding to launch the system commercially, and not to launch it *before* Super Phénix has even got off the ground, a timetable that would offend against prudence and good sense.

A premature launch of breeder reactors on a commercial scale would *a priori* reduce the chances of solar energy (and would also mortgage the chances of thermonuclear reactors, a potentially improved version of nuclear fission). To wait for the commercial breeder reactor is to offer solar energy the opportunity to demonstrate its virtues, to show that fine feathers make fine birds, thanks to an accelerated and imaginative program.

The best choice today would thus be not to choose. But can we do this? Can we be so wise as to wait? The "urgency" of nuclear energy and of breeder reactors depends on a continued increase in energy requirements, and on the progressing scarcity of fossil fuels. *In both cases it seems that the final resource, time, is not so short as feared.*

The growth of world energy demand has had its wings clipped. It is coming back to reason after an exceptional bout of fever lasting nearly a quarter of a century. This pause, before the flight into the unknown, is salutary. It remains to be seen if fossil fuels can effectively keep us going until we pause to reflect.

7

OUR LACK OF KNOWLEDGE CONCERNING OIL RESOURCES

Today the world depends on oil and natural gas for two thirds of its commercial energy. In view of the slowness and the difficulties of penetration encountered by nuclear energy and/or so-called new forms of energy such as solar or geothermal energy, it is probable that world dependence on oil and gas will still be at least of the order of 50 percent at the end of the century. In a word, it could almost be said that we are "condemned" to oil for many decades yet. Will this be possible? If so, under what conditions?

Whether we like it or not, whether we appreciate it or not, our civilization is "oil-based." There is some oil "invested" in almost everything we use. This truism goes furthest in the case of transport, and in particular the motorcar, whose characteristics and "power" have so much shaped our way of life. To move to another civilization, less oil-based and less car-minded, and more nuclear or more "solar-fusion-minded" would be a formidable change, the amplitude of which is not always appreciated. The operation is doubtless possible, but it is certainly a considerable undertaking. What should be the tempo of this change? A rush? Without having time to weigh up and test replacement options, probably the most risky and expensive way? Or should we fear that if we move too slowly we may no longer be able to achieve the join? The reply partly depends on how much longer oil will last.

For all these interrogations in fact come down to the same question: how much oil (and gas) may be produced at a reasonable price that our economies can bear within the next few decades? Or, still more briefly, how much oil is there?

The reply is disquieting and perturbing: we hardly know at all. The man who makes plans for the future or prepares for a long journey knows his financial assets and knows how much gasoline the fuel tank of his car holds. But our itinerant civilization, our oil-based civilization, does not know what capital it has, nor its potential revenue, nor how much oil it has in its tank. In our opinion this is one of the greatest paradoxes or our energy-using civilization, that it is so totally dependent on a form of energy without knowing how much longer it will last, and especially whether this will be sufficiently long.[1] Awareness of this surprising situation is recent, dating back to about the time of the oil crisis of 1973-74 (without however being exclusively generated by this oil crisis); and also, this pricking of the conscience is not universally shared, for there are many who minimize it or still do not realize the importance of the question.

This chapter will therefore be devoted to making an estimate (at least, the least-bad estimate possible, given current knowledge or lack of knowledge) of the amount of oil at our disposal. Where is it? The subject of the following chapter will be how and when can and should we produce it. It is in the light of this appreciation that the chapters on energy requirements, nuclear power, new forms of energy, the nuclear-solar debate, and the management of resources should finally be regarded.

To say that oil is found "in the OPEC countries" and to construct a policy of "replacement" on this simplistic affirmation is to see the problem only in the short term and to limit oneself to the visible part of the iceberg. The problem merits more serious and greater attention.

Before trying to answer this crucial question one may ask how we have got into this situation or, more precisely, how and why this fundamental requirement of making a reasonable estimate of our long-term oil supplies was neglected.

A first reply springs to mind: although since the beginning of the century a few experts, such as the scholar Arrhénius of the 1930s, have predicted periodically the exhaustion of oil reserves, these alarmist forecasts have regularly been defused by new discoveries, like necessary cycles of alarm and abundance. We may laugh today at these wrong forecasts, just as tomorrow today's forecasts may also be laughed at. It would nevertheless be imprudent to extrapolate this historical situation indefinitely and to deduce from it that *always plenty* of oil will be discovered.

OIL RESERVES AND RESOURCES

Because of their importance it is indispensable to understand the fundamental concepts of *reserves* and *resources,* which are often confused or misinterpreted; notions also often applied (with some minor modifications) to natural gas, coal, uranium, mineral ores, and the like, in short, to all minerals.

Reserves of oil are the estimated quantities of oil to be found in well-identified underground deposits, from which the oil may be produced or "recovered" economically now or in the near future and using techniques that are currently available or may be expected to become so in the medium term.

This fundamental notion in fact contains a number of components. It may be remarked that it allies two type of characteristics, each partially independent of the other: the state of geological knowledge on the one hand and technical and

[1] The same paradoxical situation looks like being reproduced with nuclear power and uranium resources.

economic aspects of production on the other.

- *geologic conditions*. It is a question of *estimated* quantities, and not actually measured quantities. Similarly, it is a question of oil in the ground, not yet produced. But not all the oil; only the estimated proportion recoverable.[2] The deposits containing this oil must not only be located, their *content* must be *identified* both as to position and as to extent (say with a geologic-geographic precision of estimation equal to or better than 20 percent);
- *technological and/or economic conditions:* Technological conditions concern the methods of production and/or transport. The methods used in practice depend on the characteristics of the deposit: dimensions, permeability of the rock reservoir, and the like; they determine the technical share of the cost of production. To this must be added the economic share, which includes such factors as royalties or dues, rent for land, and taxes. The most important economic factor is of course the "market price," the difference between this and the total cost of production determining the profit, that is, in many cases the motivation to produce or not.

In practice, estimates of reserves are continually changing as a result of the opposing forces of subtraction and addition. They go down naturally as a result of production, a reason for trying to "reconstitute" them in order to maintain a constant ratio of Reserves to Production (given in years). They are reconstituted or are augmented by new finds (new deposits) or by revisions or additions (to old deposits). They can also increase or decrease as a result of changes in economic or technological circumstances: higher prices tend (other things being equal, which is not always the case in a period of inflation) to increase reserves, making economic, or "commercial," deposits that were previously not so.[3] On the other hand, an increase in the cost of production, given the same selling price, tends to diminish them. Similarly, technological improvements or the introduction of new methods of recovery — like the methods of enhanced recovery, to which we shall return — can also contribute to increase reserves.

As against these *reserves,* relatively well known and hence sometimes referred to as "assured," there is a second category, *"resources"* (note that in the case of oil they are sometimes called "potential" reserves, in contrast to "assured" reserves). A distinction is made between two major categories of resources: identified non-economic (or marginal, or conditional) resources, and resources still to be discovered.

[2] Sometimes reference is made to the total quantity in situ, of OOIP (Oil Originally in Place), independent of the final rate of recovery, but this concept is less useful for our purposes. The relationship is: recoverable reserves = total quantity in situ × proportion recoverable.

[3] In fact, the substantial increase in oil reserves which was expected after the spectacular jump in price in 1973-74 did not materialize. The argument put forward is that costs have grown in parallel, and continue to do so, faster than prices (an inflationary phenomenon that OPEC complains about, too).

Identified Non-economic Resources

It is known where these are, .sometimes with as much precision as for recoverable economic reserves, but nobody knows how to produce them at an acceptable cost, and perhaps for some nobody ever will. (In this domain, as was seen in the debates of the Club of Rome, there are two schools of thought: those who think there is an acceptable technological solution to every problem and those who think there are limits to technology, natural, social and/or economic limits that in certain cases we are close to reaching already.) As examples we may quote: the small North Sea deposits, for which the installation of a costly drilling platform would not be economic; the oil shales of the Paris basin and elsewhere currently too expensive; or, in another sector, uranium from sea water: its whereabouts are known — in the oceans, on a world scale — but nobody knows how to produce it at a low price. A boundary case, that is to say, on the point of becoming commercial because of technologocal progress *and* because of an increase in price (the former permitting the latter): a part of the oil left in the earth after primary and/or secondary production (injection of water or gas) and the current goal of tertiary production techniques (or "enhanced or improved"), which are attracting growing interest.

History abounds in examples of such progress or mutations having made enormous energy or mineral resources available. One example is the exploitation of porphyry copper at a time when the United States was getting anxious about a "copper crisis" following the exhaustion of deposits in lodes, and not being able to keep up with the explosive development of the electrical industry. Off-shore drilling, which gave access to oil resources beneath the sea, thanks to progress in technology *and* an increase in prices, is another example. Then there are tar sands and oil shales, which may be exploited one day — perhaps sooner than we think,[4] and these are doubtless most interesting for the oil industry in the medium or long term. In every case there is generally a combination of the play of forces of increased prices *and* the emergence of new technology. In this area our civilization still has much to do as regards the permanent technological challenge. Concerning this, it should be noted that the industrial and developed countries here play an essential role. In most cases, like porphyry copper yesterday, tar sands and oil shales today or tomorrow, uranium from sea water the day after tomorrow (and even, in one sense, numerous future applications of solar energy depending on anticipated technological breakthroughs), the new resources thus made accessible are of a greater size than our current resources, though admittedly the quality is lower.

Resources Still To Be Discovered

These are the other unknowns in our problem. They are sometimes described as

[4] The exploitation of tar sands in Canada has already begun.

hypothetical, when they are *assumed* to be present in geological provinces already known or where oil exploration is already taking place but has not been completed. The fact that oil has already been found there is a favorable indication; thus we have some hypothetical knowledge of the resources of the North Sea. Such resources are described as *speculative* when they are assumed to be in geological provinces which are little known or unknown, such as the ocean depths (beyond 200-250 m) or the slope of the continental shelf, the Antarctic, or the Irish Sea before the first drilling commenced.

It is these latter resources (the sum of unidentified resources remaining to be discovered) that are the object of the most widely differing evaluations. However, there is not complete agreement on the others. At the time of the Delphi inquiry in 1977 into world oil resources, estimates of oil still to be discovered and produced varied between less than 60 billion to more than 800 billion tons. Penury before the end of the current century, or abundance up to the end of the next!

Note, too, that among the resources still to be discovered some will be found to be economic as soon as they have effectively been discovered (for instance, any possible large deposits in the Irish Sea) and will thus move into the category of economically recoverable reserves, which they will augment. Others will prove to be not economically recoverable within the foreseeable future (for example, with deposits that are too small and too dispersed, ''noncommercial,'' in the Irish Sea), and they will be added to the identified but non-economic resources.

These concepts, somewhat complex but indispensable, are summed up in the so-called McKelvey chart in figure 7-1, in which on a horizontal scale is indicated the state of geological knowledge and on a vertical scale the economic assessment.

It is clear that to go into or even to try to understand every coherent long-term energy policy, knowledge of *resources* (or failing this, *better* knowledge of

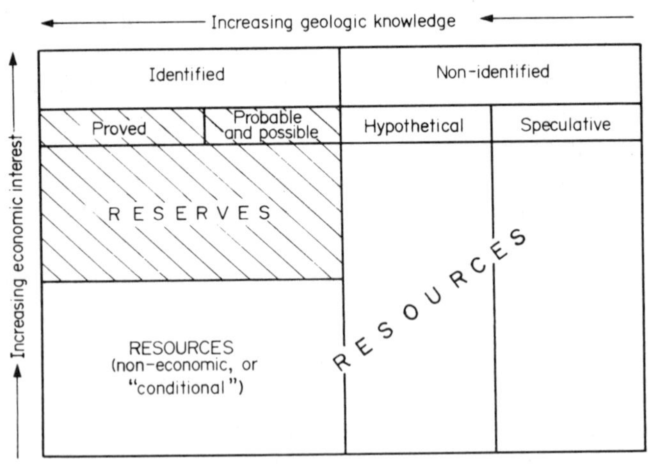

FIGURE 7-1

resources) is necessary, while short-term energy policy, on the contrary (when not kept on a short leash), is more logically interested in reserves. Knowledge of oil resources is very unsatisfactory, very insufficient, so much so that one might even sometimes ask on what basis certain fundamental political decisions were taken — whether with the blind confidence that characterized the "Belle Epoque" of oil (to use Shell's expression for the period preceding 1973) or under the pressure of irrational alarmism.

Concerning this, and to show that the situation does not only apply to oil, we may recall as an example the French nuclear program of 1973-74. Assuming installed nuclear power of nearly 200 GW(e) in France in the year 2000, it "postulated" (estimating a requirement of 5000 tons of uranium per unit of 1 GW(e) = 1000 MW(e), over a lifetime of 25-30 years) a need for about a million tons of uranium, equivalent to *world reserves* at this time, or twenty times French reserves. As the uranium *resources* were completely unknown at the time this was taking a big gamble or showed great unawareness, which makes one wonder. France was not alone in this by any means, with many countries taking a gamble on uranium that they refused to take — then or now — on oil, although the risk there may be smaller ultimately (though the alarm was more recent!).

FROM KNOWLEDGE OF RESERVES TO LACK OF KNOWLEDGE OF RESOURCES

The distinction between reserves and resources is not limited to the geological and economic factors that have lead to the McKelvey classification. There are other differences (see Table 7-1).

The cause of our relative ignorance and hence our anxiety resides historically in the big difference in interest in the two categories. Interest in reserves has always been lively. It was encouraged by economic growth, by the necessity of keeping these reserves at a certain level, that is to say, of reconstituting them as they were used up by production. Reserves are the treasury of fuel for an energy enterprise. On the other hand, interest in long-term resources (capital) was only episodic until recent years, when there has been increasing awareness both of the necessity of thinking about long-term considerations (because of the inertia of the energy system) and the non-automatic addition of new reserves ad infinitum, which had previously been assumed. This leads us to consider the time-scales for these two categories: ten to thirty years for reserves, fifty to a hundred years for resources. In other words the important ratio of reserves to production can vary between ten and thirty years, according to circumstances; if it is higher, as it is for certain minerals, this is obviously a good thing, but in general ten, twenty or thirty years represents the time-scale for industrial planning. For resources fifty or a hundred years or more are necessary for long-term national planning. If we get involved with nuclear power, will there be enough uranium in fifty years or in a hundred years, or shall we have to look for another solution, such as thermonuclear fusion or solar

TABLE 7-1. Comparison of Reserves and Resources

	Reserves	Resources
Interest	great	little in the past. now beginning to emerge
Time-scale	10-30 years	long or very long-term (> 30, 50, 100 years)
Economic	must be profitable	not profitable today, possibly appealing with high ''science-fiction'' technology
Estimates by	the industry	industry members (a lobby) or institutions (governments, universities
Data	relatively little, conservative, ''secret'', and oriented to exploration in the near future	uncertain and speculative, more oriented to obtaining scientific knowledge
Methods for obtaining data	industrial approach (expensive) exploration drilling measurement	on paper, desk research (historical statistical methods, or geological analogies)

energy?[5] In practice, leaving aside religious or metaphysical considerations, man has rarely concerned himself with such a distant horizon. It is an important achievement of our generation that it is showing concern for the generations to follow. Previously we implicitly followed the philosophy attributed to Louis XV: ''Things will last as long as I shall,'' which is not necessarily true of energy supplies.

For the name of reserves to be merited, exploitation must be accompanied by expectations of profit. Further off, resources are unknown (and profit uncertain) or, for certain people and by definition not economic under current conditions or in the short term. This does not provide much motivation.

The crucial question is to know who is interested in reserves and resources. Who is interested in reserves and makes estimates? Generally it is industry, for it exploits them and lives from them. Who is interested in resources? Nobody, or hardly anybody; just a few curious individuals or a few academic institutions. Governments? Generally not, or not yet, with a few exceptions. The American government, for example has launched several programs for long-term evaluation of resources of oil, natural gas, uranium, and certain other strategic minerals.

Other governments, such as those of European countries, sporadically manifest

[5] But in this case, is it worth getting involved with nuclear energy just for a transitional period?

dawning interest, often timidly (to judge by the budgets allocated in each case). Recently, industry, the big international oil industry, has started making estimates of the future resources of oil in order to get a picture of its own long-term future. Certain companies — happily, not all! — have even made such pessimistic estimates that they are beginning to diversify vigorously outside the oil sector[6] and even sometimes outside the energy sector (as Mobil Oil, for example).

There is no more urgent task for governments than to improve our knowledge of oil resources. To improve our knowledge of oil resources is to transfer those gradually to the category of "pre-reserves," or possible or probable reserves, perhaps less well known than assured reserves but much better known than resources currently. For resources that are identified but not economic, a scientific research program must be given priority, and not in a forced artificial manner. Evidence that programs have not been taken up on the necessary scale, and the necessity of the task, has recently been seen in the case of oil shales in France. A preliminary and worthy effort has yielded interesting conclusions, although pessimistic and much too timid. But efforts have been stopped when the scope should have been enlarged: there are several billion tons of oil waiting under French soil.

As regards non-identified sources, geological research and geophysical exploration should increase by a degree of magnitude at least, including drillings. This presupposes important changes. One of the principal ones is that an appreciable mass of information relating to geology and geophysics is being kept by the world oil industry to itself. It spends large sums of money on this "capital" to acquire ever dearer permits for exploration. It is therefore vital to rethink the role of oil companies in exploration. This is in fact happening, due to the emergence of, and pressure from, new national companies in the producing and/or consuming countries. The solution must be to seek to maximize the means for increasing our knowledge of resources and reserves and our chances of producing them according to our needs and when needed represent a world problem going far beyond the strategies of "simple" companies. On the other hand, national pride — or the secret desire for "revenge" — should not conceal the *enormous* know-how patiently acquired by the world's oil industry: know-how that today is irreplaceable and probably will be irreplaceable in the decades to come. One may be concerned to see it so underestimated, and sometimes even underutilized, while the spectre of a dearth of oil is invoked. It is an urgent responsibility of governments to remedy this situation.

This problem of knowledge of oil resources is a world problem and an international problem. It interests every government. Has not every one asked the question as to what happens after oil, the question of substitute energies? It is clear than an appreciation backed by better knowledge of world oil resources would permit better-informed decisions, not to speak of a healthier judgement of the

[6] A curious example of an industry, the most powerful in the world (but perhaps it is because of that), lowering its arms and recognizing that it is at the end of its resources in the literal sense.

CONCERNING BARRELS AND TONS

The unit of measurement used by the world's oil industry is the barrel (abbreviated to bbl). In European countries the unit used is the ton. Japan uses the cubic metre.

The difficulty of *precise* conversion of one unit to the other arises from the fact that the barrel is a unit of volume (easily measured with a meter, as on a gasoline pump or at the head of a well) while the ton is a unit of weight. The number of barrels (of 159 litres or dm^3) in a ton of oil depends, however, on its density, and each oil is different in this respect. Thus a ton of Algerian (light) oil is nearly eight barrels, while a ton of Venezuelan (heavy) oil is only seven.

To make a rapid conversion, the approximation used is 7.30 barrels to the ton. Incidentally, this is the figure valid for Iranian oil.

Another difficulty is that the world's oil industry generally publishes production figures in barrels per day, which is logical because production may vary from one day to another. In Europe tons per year are preferred. With the same reservations as above, for a quick conversion one may take 1 barrel/day = 50 tons/year (or, multiply the production of barrels per day by 50 to obtain production in tons per year; example: 100,000 barrels/day = 5 million tons/year).

As regards world oil prices, these are quoted in dollars per barrel or $/bbl, a sign with which an oilman is familiar practically from birth.

international situation in the knowledge of where the oil is, or where it is not, and who can produce it. Let us not forget that an important factor in the decisions of OPEC — decisions whose constant probable importance have been dramatized by studies by the CIA and the Workshop on Alternative Energy Strategies — is that if world oil resources were better known, some of OPEC's attitudes would probably have to be modified.

SOME FIGURES

In the absence of a specialized world organization playing on a planetary scale a role analogous to that of the American Petroleum Institute as regards American oil reserves, estimates of national and world oil reserves are published annually in the

CLASSIFICATION OF CRUDE OILS AND THE API SCALE

Professional practice is to classify crude oils in terms of density, an important parameter linked to their content of light or heavy fractions, which have an influence on their transport and refining characteristics.

On the international crude oil market the API (American Petroleum Institute) degree is generally used, related to the density or specific weight as

$$\text{Degree API} = \frac{141.5}{\text{specific weight}} - 131.5$$

The lighter a crude oil is, the higher is its degree API. For a specific weight of 0.653 one has 85° API; for a medium density of 0.850 one has 35° API; and for a specific weight of 1, API would be 10°.

Generally, special well-defined crudes are taken as reference values, such as East Texas 38° API for the Gulf of Mexico, Ras Tanura 34° API for the Persian Gulf, or Tia Juana light 31° API for Venezuela, where prices are "posted."

The variations of density and sulphur content (because of the highly polluting character of compounds with this metalloid) have been taking on increasing importance in recent years and have led to complex price differentials arising, above and below the average value. The basic price equivalence of two crude oils of different origin and with different API ratings cannot escape the accusation of being somewhat arbitrary.

technical press, principally *World Oil* and *Oil and Gas Journal*. Their respective estimates can occasionally vary by more than 20-30 percent, especially for countries not publishing official statistics. Certain major companies such as Exxon have their own statistical service.

On 1 January 1979 (according to *Oil and Gas Journal*) total world reserves of oil — relatively constant over several years, implying annual discoveries approximately compensating for consumption — were of the order of 90 billion tons. Of these reserves 57 percent are in the Middle East, and 23 percent of the world total (nearly a quarter) in Saudi Arabia (in fact, 41 percent of the world total is situated in the Arabian peninsula). This unique situation has been discussed time and time again, the fruit of the capriciousness of the Goddess Geology and man, attracted, dazzled by cheap oil and fabulous profits. We shall not add to this discussion: it is the *future* which interests us.

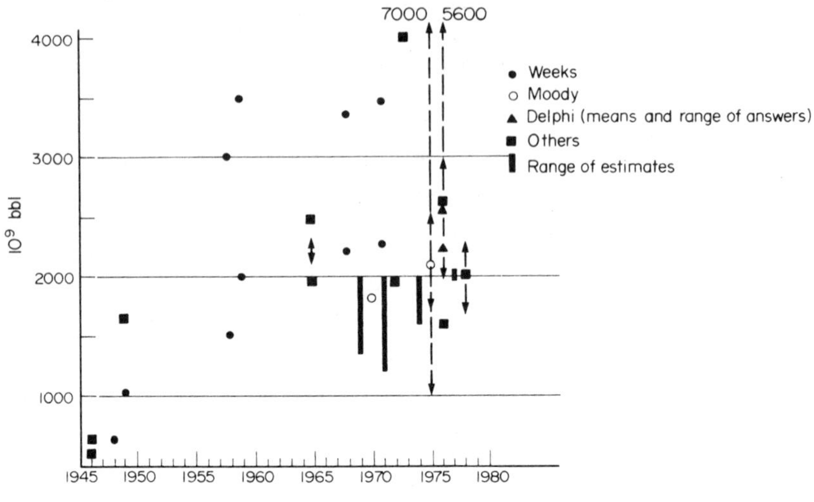

FIGURE 7-2

Estimates of world oil recoverable *resources* are much less systematic. About two dozen have been made in the course of the last twenty five years. The values — sometimes corrected to make them mutually consistent — are shown in figure 7-2. They are very interesting but call for much comment. The extreme right represents the envelope of results from a Delphi poll[7] carried out by Pierre Desprairies, president of the Counseil de l'Institut Français de Pétrole on behalf of the World Energy Conference (results presented at the tenth session of the WEC in September 1977 in Istanbul).

Some thirty experts, speaking for themselves or in the name of their organizations, replied to this inquiry. No doubt they represent the flower of their profession and combine between them the maximum amount of current knowledge concerning oil resources. Nevertheless, what differing opinions they held! The most pessimistic estimated that there are 170 billion tons in the world waiting to be produced, including those in difficult polar or deep off-shore regions; the most optimistic estimated between 550 and 950 billion tons waiting to be discovered and/or produced. In the first case, this would represent about 60 years of consumption at the current rate (not negligible, in any case); and in the latter, between a hundred and eighty and three hundred and twenty years at the current rate of production/consumption.

Before analysing some of these results, we should recall that the questionnaire, prepared with the greatest care and very precise (and improved between the first and second stage) , which should lead to a certain consistency of replies, had set

[7] A Delphi poll is generally made in two stages. In the first, carefully selected experts reply to a questionnaire. In the second stage they are shown anonymously the replies and are invited to review, modify, or confirm their opinions.

two limits: the year 2000 for discoveries and a *cost* of production of $20/bbl (compared to a current price of $12-13/bbl, all figures being in 1976 dollars). Extension of these two limits, for instance, allowing costs to go above $20, which seems probable between now and 2000, would or could considerably increase some of the figures.

Generally giving more weight to "moderate" replies, which we think are largely the replies from the oil companies, the author of the poll puts forward the following major conclusions:

> Final world recoverable resources of oil *remaining to be produced*[8] in 1977, assuming the current rate of recovery of 25 percent were increased to 40 percent by the end of the century, have been assessed by the 28 experts taken together at 260 billion TOE, not counting oil from the ocean depths and polar zones which is still regarded as unconventional, and 300 billion tons if this is counted.
>
> Limiting the analysis of replies to conventional oil (260 billion tons) three differing opinions are given by experts:
>
> —the majority (18 replies, or two-thirds of the replies) estimate recoverable resources at 240 billion tons;
> —an optimistic opinion (7 replies, or a quarter of the replies) puts the figure at 350 billion tons or more;
> —a pessimistic opinion (2 replies, or 7 percent) is 175 billion tons.

Let us look in detail at the statistical split of these replies concerning total resources, including deep off-shore oil and the polar zones: two replies between 150 and 200 billion tons, five between 200 and 250 billion tons, twelve between 250 and 300 billion tons — the bracket preferred by the oil companies, whose opinions are perhaps not completely independent, in any case — two replies of 350 billion tons, one reply between 300 and 500 billion tons, three replies of 500 billion tons, and the last reply (not broken down by regions) giving a very wide bracket of 550 to 950 billion tons. The spread of results is unfortunately significant, and the "averages" that might be derived from it must obviously be treated with caution. However, because of the frequency of replies between 200 and 300 billion tons, the average for this range (about 260 billion tons altogether, or 240 without deep sea and polar zones, as stated above) differs by only a little more than 10 percent from the overall averages (about 300 billion tons) while the average of the seven higher estimates is of the order of 480 billion tons.

It is interesting to note that these 300 billion tons of oil represent about a hundred years of consumption at the world's current level (3 billion tons per year) or sixty years of consumption at a rate doubling between now and the end of the century to reach 6 billion tons per year. Six billion tons of oil per year (not including natural gas, which could represent 50 percent or more additionally), equivalent to 9 billion

[8] This does not include quantities already produced, about 50 billion tons, which are generally included in the figure for *total* resources.

tons of coal, is more than the total world energy consumption today. This estimate of 300 billion tons, which certain experts, including ourselves, regard as conservative, gives a first idea, even if imperfect, of our oil "wealth." It acquires added weight if one takes into account the number of countries interested in consuming this oil and the number of countries interested in producing it! (One should by no means generalize the, fortunately exceptional, case of the Arabian peninsula).

The author of the Delphi poll adds: "The opinions expressed on the deep sea and the polar regions are pessimistic or uncertain, with an average of 13 percent of the 300 billion tons of ultimate resources." In fact, to judge by our own discussions with numerous experts, some of whom had replied to the Delphi questionnaire, the replies were "pessimistic or uncertain" for the year 2000, or with the ceiling price of $20 per barrel. If one sets aside one or the other limit, or *a fortiori* both of them, the potential could appear appreciably greater than the values given. These are speculative resources, as we have already said, for which an accelerated research program would help to remove the doubts in a few years. According to some experts, such as Hollis Hedberg of Princeton, this task is most urgent.

Although Pierre Desprairies insists on the interplay of the two factors, the discovery of new deposits and continued improvements to the rate of recovery (which we shall analyse in the next chapter), it is not easy quantitatively to define their respective roles. Taking into account some 90 billion tons of reserves already discovered (which would become about 140 billion tons in practice if the rate of recovery went up progressively to 40 percent), it is generally admitted that as a first approximation it would be necessary to discover about the same amount of oil over the next twenty or thirty years as has been discovered since the start of the world oil industry. The pessimists say this is impossible, giving as one reason that a second Persian Gulf will not be found. The optimists think that considerably more will be found, even if only by going over the planet with a fine comb (which would certainly take more than twenty or thirty years).

CONCERNING A POSSIBLE GEOGRAPHIC DIVISION

It will be understood that the uncertainties concerning global figures make us particularly cautious when it comes to regional figures. In fact, it is well known that uncertainties of detail may more or less cancel out when added. Apparently this is what happened with the WEC's Delphi poll.

Table 7-2 shows the geographical division representing average values from Delphi. One can see how big our uncertainty or our lack of knowledge is if one knows that the value of 28.5 billion tons the United States and Canada have yet to produce is an "average" of 27 values of which the lowest is only 6.12 billion tons (hardly higher than today's assured reserves, which would suggest that practically no more oil remains to be found in the United States,[9] and the biggest is 50 billion

[9] While exploration, after a long phase of depression, is now in full swing again.

TABLE 7-2. Comparison of Estimates from the Delphi Poll (1977) and John Moody (1975) of Ultimate Quantities of World Oil Still Available for Production (in Billion Tons)

	Delphi Poll (1977)	J. Moody (1975)
Socialist countries	59.4	58.4
United States and Canada	28.5	33.1
Middle East and North Africa	109.1	89.8
Africa south of the Sahara	11.3	9.3
Western Europe	11.2	11.1
Latin America	22.9	16.6
South and South-East Asia	15.1	16.3
TOTAL	275.5	234.6
Deep off-shore and polar zones	38.7	
GRAND TOTAL	296.2	

tons: even if one takes only the estimates with an overall total between 200 and 300 billion tons — a range of values that seems to represent a certain consensus at the global level — the values still range between 15.6 and 45 billion tons, a factor of 3, despite the fact that the North American continent is the most drilled in the world, with more than 2,800,000 holes!.

Curiously, this spread of estimates is less wide for the socialist countries,[10] which do not publish statistics, than it is for North America, which publishes the most. In fact, this apparent convergence for the socialist countries may result from the fact that the majority of experts have no direct experience of exploration or production in the socialist countries and are leaning on a few rare analyses by western experts (including the CIA).

The variation of estimates is also significant for the other regions:

- For Latin America the extremes are 7.9 and 55 billion tons if one takes all 27 replies, and 12.5 and 36.9 billion tons for the eighteen central replies (again a factor of three).
- For the Middle East and North Africa the situation is almost the same, with 54.8 and 300 billion tons as the extreme values taking all replies, and 76 and 156 billion tons (a factor of 2) taking the central group.
- The biggest differences are for deep off-shore and the polar zones, 0 to 230 billion tons for all replies and 0 to 50 billion tons for the central group. Remember the remark that the value of 0 might apply because of the time limit imposed — the year 2000 — or the stipulated cost of $20/bbl.

[10] The extreme values are 27.3 and 96.3 billion tons if one takes all the replies, and 40.3 and 83 billion tons if one takes only the "moderate" values (for all the socialist countries such as the USSR, Eastern Europe, and China).

In the same table we have given the values put forward by John Moody at the World Oil Congress in Tokyo in 1975. There seems to be reasonable agreement between the estimates. In fact, the average values given by Moody, despite representing the unique views of one forecaster, conceal the same uncertainties as the Delphi poll values, reflecting the different opinions of many experts: for the United States and Canada his bracket is from 12.3 to 35.5 billion tons of oil still remaining to be discovered (again a factor of 3). John Moody was one of the pioneers of detailed analysis of basins and oil-bearing areas (''plays''). His research was initially confidential to his company, Mobil Oil, and was made for strategic purposes of world exploration. Moody succeeded in publishing his figures before the apparently much more optimistic figures published by the U.S. Geological Survey at the beginning of the 1970s. The polemics Moody's figures provoked undoubtedly stimulated world interest in this problem, as is seen in the Delphi poll and other recent studies.

Table 7-2 once more reflects the considerable importance of the Middle East and North Africa, although their relative shares are considerably reduced (about 37 percent of final resources according to the Delphi inquiry, as against 62 percent of reserves on 1 January 1978). The socialist countries come a good second.

Indeed, without seeking to minimize the exceptional importance of the Middle East, there are certain experts who dispute, on the other hand, the inadequate shares attributed to other regions, pointing out that as regards the state of our geological knowledge, or lack of it, world-scale exploration is still in its infancy and there is an enormous disproportion in the geographical spread of drilling operations.

FROM GRUYÈRE CHEESE TO THE ISOLATED HOLE

Some sceptics base their criticism of these results on the state of exploration and drilling. Geophysics is a powerful tool to help indicate at a given moment in time, that is, with the current (limited) state of our knowledge, where to find the most promising places to drill, for it is necessary to drill. Unfortunately, even today, only drilling can remove doubts and show whether there is oil in this marvellous structure that has been spotted — as a reward or as a little bit of luck[11] — or sometimes it is a dry well, and one must look further afield, or elsewhere.

This spotting, this selection does not imply by any means that there is oil anywhere within this structure. The best proof of this is that as technology progresses (for example, there has been spectacular progress recently using three-dimensional seismics, in color, the method of ''bright spots'') the explorers return with success to areas previously neglected or insufficiently explored. In addition, the drilling equipment now available permits the explorer to go deeper. One of the most striking examples today is the south of Texas, one of the most drilled regions of the world, where, however, exploration has taken on a new

[11] This still happens. If not at the level of the big companies — although even this could happen — at any rate with the smaller companies or independents.

impetus. Thanks to new techniques "we are now seeing what we didn't see before." In consequence "we are now looking for things we didn't look for before" (including stratigraphic traps). Though over a hundred years old, the oil industry remains a young industry, and its potential for development and technological progress remains considerable, ranging from the possibility of changing a part of the bit at the bottom of the hole without having to bring up the drilling train, to the revolution which would guarantee the presence of oil without the expensive necessity of drilling.

Drilling remains indispensable and is the best tool available to improve our (insufficient) knowledge. The questions that spring to mind are: where was the drilling done, how was it done, and was enough drilling done?

First, the target! According to Michel Halbouty, a lucky and independent oil prospector and international expert, there are about 600 sedimentary basins in the world that bear oil or, at any rate, fulfill all the conditions to suggest they might bear oil. Of these, 200, a third, have practically never been explored, often because of their poor accessibility. Halbouty goes on to add that according to geological knowledge, as only 12 to 15 are likely to have reserves of over 1.5 to 2 billion tons (i.e., like the North Sea), unfortunately the probability of finding a new Persian Gulf is not very great. Two hundred and forty basins have been explored to a small or moderate extent but have not yet gone into production (for reasons of cost, and/or because it was not necessary). There remain 160 basins being explored, those which provide our supplies today. Of these 160 basins, 25 (about one in six) contain more than 1.5 billion tons of oil (or gas, in equivalent energy), representing about 86 percent of all hydrocarbons discovered to date (these figures, and those that follow, well illustrate the importance of the giant deposits that the oil industry, interested in profit,[12] has preferred to investigate, a phenomenon to which we shall return). Six basins from these 25 contain more than 7 billion tons of oil (or gas, in equivalent energy) each, representing together 65 percent of all hydrocarbons discovered to date. One basin, the Arabian-Persian Gulf, contains a little more than 40 percent of world reserves.

The evidence is that there is still something left in the cupboard to be looked for. Bernardo Grossling (of the U.S. Geological Survey) goes furthest, showing with figures the considerable gap between the degree of exploration according to *countries* and not just basins. Considering some 3.3 million holes made — all types of drilling, not just exploratory drilling — in the world since the beginning of the oil industry, 73 percent (nearly three quarters) of all world drilling has been done in the United States alone, and more than 95 percent in the developed countries. In one sense, it is understandable that oil was looked for where it is consumed, but it is a well-known — and irritating — fact that that is not where it is generally to be found.

[12] This is not a criticism, but just a statement of fact. "Profit" has permitted the formidable development of the oil industry, from which everybody "profits." However, this does not exclude taking another look at the problem today from a different angle.

The Middle East, of course, bears out its exceptional character: not much drilling, plenty of oil. It is interesting to compare the other regions, those of the developing world, among others, where less than 5 percent of all drilling is done. In Latin America, leaving aside Venezuela and Mexico, which have three quarters of the holes made on this subcontinent, it is evident that there is a vast disproportion in drilling activity.

Going further, a comparison may be made of these drilling activities either with regard to sedimentary areas that may contain oil or in terms of the regional estimates from the Delphi poll.

As regards drilling in prospective areas,[13] the figures for "density" of drilling vary from 370 holes per 1000 km^2 for the United States (a gruyère cheese) to 8.5 for all Latin America (but 130 holes per 1000 km^2 for the group Argentina + Mexico + Venezuela) and a little less than one hole per 1000 km^2 for the prospecting zones of Africa and Madagascar, 400 times less than in the United States.

The comparison with the estimates of the Delphi poll is no less interesting. Regard the United States: 73 percent of world drilling, for about 11 percent of final resources. This illustrates the "oil maturity" of the United States. Nevertheless, despite this "maturity," the degree of uncertainty concerning final resources, is, as we have seen, a factor of three if one takes the central values, and a factor of eight if all the estimates are included. This uncertainty relates to the extent of the areas worth prospecting in frontier regions: off-shore oil from the Atlantic, from the Pacific, from Alaska (underground and off-shore). The state of oil maturity of the United States and, to some extent, Canada does not prevent the experts attributing the most substantial relative addition to reserves. Indeed, the ratio of 28.5 billion tons of final resources to 4.8 billion tons of assured reserves, 5.9 to 1, is the highest of all the regions considered. It is only 2 to 1 for the Middle East and North Africa, and 4.15 to 1 for Latin America. This would mean that the least explored regions are the closest to their final potential. Here there is a paradox that links up with the arguments of Grossling and tends to prove that many regions have been greatly underestimated by most experts.

In absolute terms, how could a region as well explored as the United States[14] ultimately multiply its reserves by 5 or 6? Essentially by two processes: by exploring frontier regions, on the one hand (provided that the environmentalists cease to oppose this), and by going over the country with a fine comb of drilling and exploration, on the other hand.

It may be noted that, crossing the Atlantic, European countries such as France, Italy, and Spain, some of which had a deceptive attack of "Texas Fever" in the 1950s and 1960s, are slowly doing the same: the "frontier" regions (the Gulf of

[13] Estimates of prospective areas by Grossling have been challenged and reduced recently by Ivanhoe.

[14] Some people dispute this, saying that, in consideration of the fact that many of the holes are "old" and rather shallow and physically insignificant and because the frontier regions are extensive and are substantially unexplored, only 2 percent of the potential volume is really *known*.

Gascony yesterday, the Irish Sea today, the Mediterranean tomorrow) are or will be gradually explored, and the fine drilling comb is slowly going back and forth, as in the Paris basin, for instance.

PROVISIONAL CONCLUSIONS ON CONVENTIONAL OIL[15]

What conclusions can one draw from these considerations? The first conclusion is that a lot, if not everything, still remains to be done to acquire a better knowledge of the final quantities of oil. Perhaps they will never be known, or perhaps only after having extracted what will have been called "the last" drop from "the last producing well" — or at least reasonably assured quantities for the future in the medium and long term. *Only such better knowledge would permit intelligent and reasonable planning for substitution by new forms of energy* and a partial extension to the era of oil by having recourse to nonconventional oil.

As a second conclusion, might a value be proposed as a basis for reflection or as a working figure, until a better one turns up? While recognizing its fragility, we may say that the average value of the Delphi inquiry, 300 billion tons, including deep off-shore and the polar regions, is the least bad estimate currently. It is a significant amount of oil, even if it is manipulated!

To put these 300 billion tons in perspective we may quickly analyse the probabilities that this figure is too high and must be revised downwards, or, on the contrary, that it is too low, and could be revised upwards.

POSSIBILITIES OF REVISION DOWNWARDS

1. The recovery rates (which we shall discuss in the next chapter) could deteriorate, which is the belief of people thinking that more or less all the good deposits have been discovered. At the same time, the hope of pushing up the recovery rate on a world scale to 40 percent (not to mention 60 percent, a figure still mooted by Soviet experts) might not materialize.
2. The costs of exploration and production, which have increased strongly in recent years, not only in off-shore drilling and in zones difficult of access,[16] might continue to grow. It is partly because of the big increase in these costs (inflation) that reserves, contrary to what many people hoped (the economic law relating reserves to price) have hardly been revised upwards, despite the rise of the price of oil since 1973 (seven times higher in 1979 than in 1973).
3. The potential of the Middle East and Western Siberia (which if geographically united really constitute the big "fertile crescent" of oil basins) has perhaps been overestimated.
4. The climate of international mistrust between producing countries, consum-

[15] We are using "conventional oil" here in the usual sense of the word, as opposed to "nonconventional oil" discussed in the next section. P. Desprairies excludes deep off-shore and polar zones from the conventional oil category.

[16] But it may be commented that this phenomenon will probably affect all forms of energy.

ing governments, and big international or national oil companies, together with short-sighted tax policies (not to speak of the threats of divestiture of the big oil companies), might discourage oil companies, leading them to shift their efforts to areas that, if not more profitable, are at least much more "quiet." This would create a situation in which competence and know-how would be lacking at a time when finding and producing oil will be becoming more and more difficult.

POSSIBILITIES OF REVISION UPWARDS

1. The discovery of a new Middle East (or even a new Western Siberia) adding, "at a stroke," some 50 to 200 billion tons or more to reserves. The chances of the existence of such a find seem slight today, but cannot be completely dismissed. Recent events in Mexico give some support to such hopes.
2. Deep off-shore — the slope of the continental shelf, and enclosed ocean basins — still poorly known. This is very evident in the big spread of estimates in the Delphi pool. Although some disillusionment has recently become prevalent, less than a hundred dry holes for millions of km^2 are hardly significant, and there are still good grounds for hope.
3. Occurrence of new types of deposit. Up till now more than 95 percent of deposits have been found in anticlines,[17] the theory of which was put forward over a hundred years ago. Stratigraphic traps,[18] which are now successfully being sought, have to date yielded only small deposits (with the exception of Texas East).
4. Improvement in oil drilling techniques on land (a possible change in techniques greatly reducing costs) permitting increasing exploitation of smaller and smaller deposits. It should be remembered that fields with less than 1 or 2 million tons of reserves account for 15 percent of United States production and that this share will probably continue to grow. The proximity of consumers that justifies the exploitation of such fields is a phenomenon that is becoming more general, even in the developing countries with a concomitant shift in interest from international to national companies. The World Bank, which recently decided to aid certain developing countries extending their oil exploration, is in favor of this evolution.

   In the same spirit, improvement or change affecting off-shore production (the result of new experience, as with the Castellon field near the Spanish coast of the Gulf of Valencia, developed by Shell, which is experimenting with a mobile installation mounted on a tanker instead of the costly platforms usually used) or technological fall-out from deep off-shore research could open up a number of fields today classified as noncommercial.

[17] Mountainous folds with an enclosed structure favorable for the retention of oil.

[18] Discontinuities in sedimentary layers.

These developments tend, when taken together, to counterbalance the increase in costs previously mentioned above.

5. Final resources may well have been underestimated, as Bernardo Grossling seems to indicate, probably with excessive optimism.
6. Techniques of enhanced recovery may also come up to expectations, or exceed them. Remember that an improvement of one percentage point in the recovery rate can mean 6 to 10 billion tons of oil. Certain techniques are still in their infancy, and are capable of making real progress if backed by a more scientific approach. The ball is in the court of the laboratories, prior to being returned to the ground.

Non-conventional oils, or new forms of energy from oil

It is easy to be ironical about "new forms of energy from oil," saying that they were first announced more than fifty years ago and are being announced still for tomorrow, and they always turn out to be 3 to 5 dollars dearer than conventional oil, whatever conventional oil costs. Indeed, it must be understood that these are expensive forms of oil, and difficult ones. But they are *abundant*. To date there has been little or no reason to produce them on a large scale. Their development has practically never been promoted, if not simply discouraged. Let us only hope that Canadian decisions to exploit tar sands and the ambitious synthetic fuel program of President Carter will really be pursued. It looks as if these are taking off at last.

By nonconventional forms of oil are meant — ranked in order of difficulty of working — heavy oils, tar sands, and oil shales. The heavy oils (density between 10° and 25° API) are thick and very viscous, hardly flow, and cannot be produced using traditional methods. Tar sands (density between 7° and 10° API) are huge sand deposits, generally shallow and containing a very thick oil, tar, or bitumen. The oil shales — which are not in fact shales in the strict geological sense of the word — contain solid matter called kerogen (with 30 to 250 litres per ton being the fraction it is hoped is exploitable), which produces a heavy oil by distillation. In this chapter we shall limit ourselves to some considerations concerning the size of these resources, remembering that a resource can be defined on two axes (the McKelvey diagram) in terms of geological knowledge and economic potential, the latter depending closely on techniques of production and the market price.

The first important point is that such deposits have been "found" but have practically never been "looked for," because of the lack of interest already mentioned. The result is that the knowledge available on a world scale is still way behind knowledge of conventional oil deposits, which itself leaves a lot to be desired. Nevertheless, despite this skimpy knowledge, those deposits recorded probably contain much more oil than the richest conventional oil fields. In fact, the super-giant oil deposits, like Ghawar in Saudi Arabia (10 billion tons) and Burghan in Kuwait, look like small deposits if they are compared to the principal deposits of heavy oil, tar sands, or oil shales, just as they are in comparison to the biggest coal fields (of several hundred billion tons).

THE GENESIS OF HYDROCARBON DEPOSITS

According to present-day theories, the environments in which source-rocks (or organogenous sediments) have been created capable of producing oil are various, including marine areas (lagoons, deltas, coastal zones, confined basins, continental shelves) as well as aquatic land areas (lakes and marshes). The limitation of microbiological activity in the sediments in the course of formation preserves part of the organic matter from mineralization, that is, its decomposition principally into carbon dioxide, water, and nitrogen. The organic matter fossilizes and in part remains a captive of the sediments: that is *kerogen*. Certain rocks, like boghead coal, contain large quantities of kerogen, up to 40-50 percent by weight, and are actually burnable rock. Most often kerogen only represents a fraction of the order of a few percent by weight. These rocks are found over wide areas and contain a considerable amount of organic matter, but doubtless most of them will never be utilizable because of their too-low concentrations (for instance, it would be necessary to use more energy to win the oil than the quantity recovered itself contains). Between these two extremes come the various oil shales, capable of providing between 25 and 250 kg of oil per ton, or at least between 25 and 150 kg per ton.

Most often, oil shales and their kerogen have not remained close to the surface but, on the contrary, have gradually become buried under increasingly thick layers of sediments and have been subjected to the action of pressure and temperature (an effect of the geothermal gradient, generally 3° C per 100 m, but exceptionally reaching 10° C per 100 m). As they bury themselves underground, the oil shales go through the "oil window" between 60° C and 135-150° C, in which "petroleogenesis" of kerogen occurs (and of minute quantities of liquid protopetroleum that has already been formed). Beyond 135-150° C, as the process of burial continues, so does decomposition, leading to the formation of lighter products, natural gas, methane, and graphite; this explains why practically no oil is found below a certain depth (contrary to past hopes) but more and more often gas.

Generally the liquid petroleum formed does not remain in the source rock. If this rock has even been in contact with other permeable rocks, the oil will have migrated, either to the surface, where it is lost (seepages or oil springs), or to a "trap" or reservoir closed by an impermeable rock of suitable structure such as an anticline or a tectonic discontinuity. It is in these that oil is sought, and where it is most often found. It can also be trapped in stratigraphic discontinuities, which are beginning to be prospected today with the help of more refined seismic methods.

However, the oil does not always quietly await the prospector's drill. It may have an interface with surface water rich in microbes, or the sediments may be lifted to orogenesis (or the formation of mountains), bringing the deposit closer to the surface, and the oil may be subjected to scrubbing, oxidation or biodegradation (beginning with the n-alcanes, or saturated hydrocarbons, then iso-alcanes with branching, and so on). The result of this alteration is heavy oils (of density between 10° and 25° API) and tar sands (of density between 7° and 10° API), which are much too viscous to flow naturally.

In fact, there is remarkable continuity between these resources, which were previously considered completely distinct.

Scientific understanding of hydrocarbon deposits of all sorts has made big progress in recent years, thanks to a "unitary" genetic theory (see box), which shows a link between resources that previously had been considered completely different. Among other useful spin-off benefits, this better understanding should also aid the "new" exploration for nonconventional oil as well as for conventional oil.

In the course of drilling in search of conventional oil, oilmen have often come across an accumulation of heavy oil. Such accumulations tend to be classified among resources that are identified but not economic, because of their too-high cost and the absence of tried and tested technology. They are "noncommercial" except in California, where they contribute about half of local production (and in other countries such as Venezuela, but on a small scale). Today or, better still, tomorrow, with higher market prices and the emergence of enhanced methods of recovery (steam treatment, for example) there will be a return to these deposits, which have the advantage of having already been discovered, though no inventory of them has even been taken.

The most famous deposits of heavy oil, and the most important today, are situated in Venezuela, in the Orinoco Tar Belt, which is not really a belt, or tar. The resources in the ground are estimated at 100 to 300 billion tons, and sometimes even figures twice as high have been put forward. Taking a rate of recovery of 30 percent or more, one arrives at a figure of between 30 to 90 billion tons of recoverable oil. This oil would not be too expensive to produce, but is of very poor quality (3 to 5 percent sulphur, and a high content of nickel and vanadium) and must be pretreated or upgraded to reach a quality comparable to the crude normally supplied to refineries. According to the experts, the cost of this improved crude oil would be between 12 and 20 dollars per barrel, but probably closer to 12 than 20 (1978 dollars). These heavy oils constitute a nice pear with which to quench our thirst for oil.

The tar sand deposits in Alberta, Canada, where commercial exploitation is only just beginning, contain more than 90 billion tons of oil *in situ*. About a third, more

Table 7-3. World Resources of Oil Shale (Donnel, 1976)
(in billions of tons of oil)

	Near to possible commercialization	Non-economic
Africa	1.5	13
Asia	3.5	12
Europe	4	7
North America	11	300
South America	7	110
Total	27	442

than 30 billion tons (three times the size of the super-giant Ghawar in Saudi Arabia) are thought to be recoverable, some by quarrying and the rest by *in situ* methods in the course of development.

The seven most important heavy oil and tar sand deposits currently known contain as much oil *in situ* as the 260 largest conventional oil fields in the world. But the record for oil potential is doubtless held by the oil shales (see table 7-3). Lying relatively close to the surface are quite a number of oil shales, of which few have been recorded or had their capacity estimated. Only a very small number of them have been the subject of studies and/or projects, let alone put into production. The principal producing sites today are in the USSR (Lithuania) and China, not to mention the ''historical'' French site of Autun. The principal projects are in the United States and Brazil. According to Donnel of the U.S. Geological Survey, recoverable resources in conditions close to being economic are currently of the order of 30 billion tons, which is by no means negligible but nevertheless presupposes a deliberate policy. Non-economic resources in the ground are estimated at over 400 billion tons, including more than 300 billion in the United States in the Green River formation, which spreads over a part of Colorado, Wyoming, and Utah, covering more than 40,000 km^2, the layer being about 20 m thick, with a content of over 120 litres of oil per ton of rock.

CONCLUSION: OLD FIDDLES STILL PLAY GOOD TUNES

We are not saying in conclusion that by adding all these oil resources there will be enough to last till the end of time. For a start, this would be wrong. And secondly, this would just be another example of that immoderateness that has so handicapped better understanding of the energy problem.

Our conclusion will be a confession of ignorance, but also a counsel of prudence. Until very recently, much too little was done to arrive at an understanding of all forms of oil resources. What was done was from a certain limited viewpoint. The most reliable figures are undeniably conservative, but how conservative?

However conservative, they are not small: 250 to 300 billion tons, that is eighty to a hundred years of production and consumption at the current world rate. Nor should we forget that in principle this is oil at *less* than 20 dollars per barrel (1976 dollars), a price that other sources of energy cannot be sure of undercutting, when all is considered.

In reality it is clear that numerous decisions have been taken only with an eye to reserves of oil, and not to potential resources, which for a long time could replenish these reserves, gnawed at by increasing production.

Sometimes the question is asked: even if there were 50 percent more, 100 percent more, would that change anything, because at any rate it will be necessary to do without it some day? And also, is it not "ethically" desirable to keep it for privileged uses? (We shall later return to this point.) In our opinion, fifteen or twenty years more to effect and master the transition to the post-oil era can change a lot of things. Within ten years we should have reached an important stage, perhaps a decisive one, in the development of thermonuclear fusion. Studies into solar energy, now mostly in their initial stages, by then will be beginning to bear fruit. This is to say nothing of nuclear energy, where a lot remains to be done.

Also, it may be thought that better knowledge of world oil resources could validly modify the opinion of certain producing countries, currently alarmed by the threat of too-rapid exhaustion of their reserves and even disconcerted by a certain "obligation" to produce. The replacement value of oil, which rightly plays a growing part, could also be better appreciated.

Finally, we are sometimes told: if governments, or the electorate, are persuaded that there is more oil, no decision will be taken to prepare for what will take place when there really is no more! This is a dangerous argument, and a two-edged one. For if there really is more oil than is generally supposed, is it not just as urgent not to "divorce" ourselves from it too soon, to convert prematurely to less well-known forms of energy that have been insufficiently studied and are insufficiently tried and tested? It is perhaps by losing belief in oil too soon that governments are beginning to set off the real oil crisis. In fact, it is indispensable to obtain better knowledge of oil resources, as cannot be repeated too often.

On the basis of the figures available, and while waiting for better ones, what scenarios of production can be imagined?

8

OIL: ABUNDANCE OR SHORT SUPPLY?

The oil in the ground we hope to recover in the short or long term is the indispensable capital we need to know about before we can draw drafts on the future. But it is also necessary that this oil be produced — if possible, intelligently and on a world scale, on the scale of human needs that are continuously developing.

Oil production evidently depends on various physical, geological, technical, economic, and political factors. There are numerous recent studies by the Workshop on Alternative Energy Strategies by Carroll Wilson of MIT, by the big oil companies such as Exxon, Texaco, Shell and BP, and even the CIA[1] whose interest in oil questions is growing. They have gone on record that world demand for oil could exceed supply within the next decade and even, according to some studies, within 1981-82, setting off a new energy crisis. This new crisis might even be more severe than that of 1973-74, because this time it would be a structural crisis.

This "publicity," which betrays often sincere anxieties, is evidently a means of bringing pressure to bear on governments and generally on all kinds of decision-makers to make them decide on a realistic and forceful energy policy. Proponents of nuclear power have no cause to allay these anxieties, which serve their purposes well.

The oil "balance" depends on two terms: supply and demand. The crisis — or the "non-crisis," or even the "crisis-perhaps-but-when?" — evidently depends on comparison of these two terms. The demand for energy was discussed in our first chapter. Suffice to note that it has doubtless been overestimated.

As regards supply, we were among the first to indicate that, on the other hand, it has doubtless been underestimated. Current events, with a market that has temporarily gone into surplus, seem to confirm our opinion. Many experts thought this situation could continue up the middle of the 1980s and even perhaps to the beginning of the 1990s.[2] Then came the Iranian crisis, again reversing the

[1] This same CIA a few years ago boasted that the only secret it could not find out was world oil reserves. It will have made some progress since then.

[2] Once more illustrating the tendency to go crazy and act irrationally where oil is involved.

situation. Will new sources of supply (Alaska, Mexico, Egypt, Malaysia) restore the balance? The energy problem remains grave and will probably continue so in the next few decades, with the permanent if latent threat of aggravation. These oscillations could be an indication to do something other than what was planned under the shock of the crisis.

CONCERNING RATES OF PRODUCTION

The first aspect of oil production is a physical phenomenon linked to the very nature of the deposits. Because of the flow properties of fluid oil (depending on its density and viscosity, and variable from one oil to another) in the porous and permeable rock reservoir (useful porosity can vary between 0 and 30 percent and permeability by a factor of more than 1000), there is an optimum rate at which a well can produce. According to conditions, if the rate is too fast, there is a risk of damaging the deposit and getting out less oil than at a more moderate rate. If on the other hand the rate of extraction is too slow, the profitability of the investment declines. Thus in the North Sea, for example, the investment is enormous (about $5000 to $10,000 is needed to create a capacity of one barrel per day[3]). Attempts are made to recover costs as quickly as possible, that is, to reach maximum production as rapidly as technical considerations permit without running the risk of damaging the deposit.

Maximum production is followed by a plateau (with constant or slightly declining production) and then a phase of decline, with a rate of, for example, 10 percent of the remaining oil being recovered annually. The "life" of a North Sea deposit can thus be fifteen or twenty years. In the Middle East, on the other hand, where the investment cost is ten to twenty times less (if not even lower) and numerous deposits have already been amortized, much slower rates of production are feasible, which is also appropriate to the size of the deposits.

In fact, the "life" of a deposit is still more complex, because production, which diminishes reserves, can be compensated for by additions to the reserves (an extension of the deposit, with better delimitation) or may change in type and rate. Primary production may utilize the natural characteristics of the deposit, while secondary production may involve injection of water or gas, to be followed possibly be tertiary production, enhanced methods of recovery, and so on. It is not, as the layman tends to think, simply a matter of digging a hole at random and counting the dollars flowing through the pipe. It is a question of managing as best as possible a reservoir that is ill-defined and subjected to phenomena that are not always well known, and this 2000 to 3000 metres below ground, in an environment that is hardly better known.

Oil is a nonrenewable prime material. Once extracted from the ground it is used

[3] In calculations of the cost of oil, it is of course necessary to add operating costs to the amortization of the investment. Because of the climatic conditions and the necessary infrastructure, these operating costs are very high in the North Sea.

and thus "lost." If reserves were fixed and known, the problem of the relationship between reserves and production would not arise because reserves would diminish in line with production. At best, some threshold of the ratio reserves/production could constitute an alarm signal: given a certain rate of production, there would be enough remaining only for so many years. In fact, there are periodically additions to reserves at the level of a deposit, a basin, a country, or of the world, that is to say, quantities of oil are transferred from *resources* to *reserves*.

MECHANISMS FOR ADDITIONS TO THE RESERVES

There are two transfer mechanisms corresponding to the two categories of resources we have considered, identified but non-economic resources and non-identified resources yet to be discovered (hypothetical or speculative). It is important to distinguish between these two mechanisms by which resources become vital reserves.

The effect of economics (such as higher prices) or technological progress, in practice often combined, permit the exploitation (that is, classification in the category of recoverable reserves) of deposits previously ranking as known but noncommercial. This is the case with a number of small American deposits, which have become, or once more become, commercial with the increase in prices. This was also the case with the North Sea (or Alaska), which could not have been put into production if the price of oil had remained at 2 or 3 dollars per barrel. It is the case, too, with numerous off-shore deposits. It is particularly the case for techniques of enhanced recovery (such as steam, *in-situ* combustion, chemical products). It is interesting to note that an *upper* limit can be put on these additional possibilities insofar as these involve already identified fields.

For assisted recovery, for example, if one takes into account the fact that already some 130 to 140 billion tons of recoverable oil have been discovered in the world (of which about 50 billion tons have already been produced) and making the two assumptions — too simplistic and incorrect, but adequate for our argument[4] — that each field was susceptible to improved recovery and that the *average* rate of recovery had been to date of the order or 25-30 percent of the oil in the ground, then one could hope to add to the currently known reserves of 90 billion tons:

- between 45 and 80 billion tons if the rate of recovery were a uniform 40 percent (an estimated possible value according to the majority of experts participating in the Delphi poll);
- between 90 and 135 billion tons (which would more than double current reserves) if the rate of recovery reached 50 percent. This without having discovered a single new deposit.

In fact, all these fields, whether having already produced or being currently in

[4] Remembering a quip of Paul Valéry: "What is simple is always wrong: what is not is unusable."

production or to be put into production in the next few decades are not and will not be equally susceptible to enhanced recovery. These figures give an idea of the importance of the techniques for enhanced recovery, which have been experimented with here and there (in California thermal methods are already systematically being employed) and which seem to be seeing the dawn of actual commercial use. Incidentally, it may be noted that the *scientific* study of these methods — including chemical, microbiological, and other methods — is largely open to the international community and especially to the industrialized and consuming countries (which do not always have oil, but which do have ideas) which may play an important role here. This study is not solely limited to the big oil companies. These, and smaller ones, of course, remain indispensable for research in the ground and ultimate commercial utilization of the processes. Here there is a possibility of evolution in the oil extraction industry that is not always appreciated.

The second mechanism of transfer to economic and technically recoverable reserves is the transfer from non-identified, hypothetical or speculative resources through exploration and the discovery of new deposits. In absolute terms, that is to say, if one ignores the time factor, it is exploration (via geologic and geophysical studies as well as exploratory drilling, which is still indispensable) and discoveries that will help us improve our knowledge of our patrimony of oil, knowledge which we have said in chapter 7 is indispensable. In reality, this exploration and these discoveries — the first unfortunately not *necessarily* leading to the second! — take place at a certain rate depending on investment, targets or explored territory, and chance. Let us not forget that the search for oil is not an exact science, and there remains in oil exploration an *inherent* risk.

THE CRUCIAL PROBLEM OF FUTURE DISCOVERY RATES

To calculate a curve for future production is rather like the problem facing a pensioner (and one knows that happy pensioners are rare), that is, to establish a relationship between expenses (production), variable capital (reserves), and income (additions to reserves, finds). Now, discoveries of oil are much more like stocks, unforeseeable and fluctuating, than stable and sure.

What will be the possible rate of oil discoveries in the next few decades? This is a major problem, widely debated. The first reaction, as usual, is to look to the past in order to try and forecast the future, in the absence of other methods (and what errors have been committed by applying this "trend" technique!). Since the early days of oil research the pattern of annual finds has been essentially irregular, marked by a few big "inventions" like Texas East in 1930, the Arabian Peninsula in the 1930s and after the war, and Alaska in 1968. To weigh up these irregularities it is usual to take averages over three or five years, which attenuates but cannot entirely conceal the essential role of discoveries in the Middle East.

The rate of annual discoveries generally depends on two factors: the amount of exploration undertaken (which itself results from the decision-making process) and the potential of the discoveries, that is to say, the quantity of resources remaining to be found (unfortunately, not known). Improved knowledge of these resources would increase the chances or reduce the risks appreciably.

As regards the amount of exploration, one may recall John Moody's more than pertinent remarks. It is well known that periodically the end of oil is announced, as it is being announced today; or like the magazine that mistakenly announced the death of Victor Hugo some twenty years too soon and when he actually died began an article by recalling that it had been the first to announce the news. Moody compares the American oil situation in 1922 and 1977 (see table 8-1). One may note the pessimism of the 1922 forecasts. Up to 1977 about twelve times as much oil was produced as had been estimated as the total known and still to be discovered; and there is still more waiting to be produced. Evidently, one may ask whether we are being equally pessimistic today.

TABLE 8-1. Two Views of the American Oil Situation.

	in 1922	in 1977
Annual production (millions of tons)	65	400
Cumulative production (billions of tons)	0.75	15.34
Remaining reserves (billions of tons)	0.68	6.8
Potential resources still to be discovered (billions of tons)	0.55	20.5

In 1922 a committee of famous oil experts convened by the American government, *convinced of the state of maturity of their knowledge of reserves and resources,* recommended the government to face the imminent oil crisis presaged by the figures and "to eliminate waste, improve the efficiency of the utilization of energy, to control its use, to extend the utilization of coal, and to carry out research into synthetic fuels based on coal, to improve the techniques for recovering crude oil and to develop an oil shale industry." Apart from solar energy, is this not more or less already President Carter's energy program? The most interesting feature is that the only thing that actually happened had not been advocated (and hardly is in the Carter program!): increased exploration and drilling to increase oil reserves and production.

Even taking the figures from the Delphi poll, generally recognized as conservative, there is still plenty of oil to be discovered. How and where is it being sought today?

The analysis of exploratory drilling in the world in recent years once again reveals the considerable importance of the United States, which on average drills 90 percent of all the wells outside the socialist countries (which updates and confirms the historical results of Grossling in the preceding chapter). What is perhaps more important is that the increase of drilling in the Western world since 1973, that is, since the oil crisis, is practically entirely due to the United States! Curiously, by mid-1977 there had been no response to the 1973 crisis by the rest of the world. This is due, or could be due, to numerous factors.

In order to drill, complex and costly equipment is necessary (derricks or rigs, offshore platforms) of which the world's stock — about 35,000 units, not counting the socialist countries — can only increase slowly, as ordered by the oil industry, which thus indicates its confidence in the future.

It is perhaps this confidence that was lacking, possible because of the "antiphon" on the end of oil in the consuming countries,[5] or because of the fear that there were not enough explorable prospects in which to utilize these devices, this being justified by the nationalistic attitude of the producing countries.

It is also worth mentioning that with an average ratio of reserves to production of at least 30 years on a world scale, the need to drill seems less urgent than in the United States, where the corresponding ratio is less than ten years and hence obliges and condemns them to unflagging drilling, high and low.

It is also interesting to note that the major increase of effort in the United States' drilling has hardly produced any increase in reserves, contrary to what one might expect or hope. The reason, not always admitted, is that this drilling has been primarily aimed at increasing production or, rather, at modifying the structure of production. Because of federal control of prices it is more interesting to develop a new deposit, where the authorized selling price is around $10 per barrel, than an old (pre-1973) deposit, where the selling price is limited to less than $6 per barrel.

In fact, a more detailed study would verify that the apparent standstill of drilling in the Western world, with the exception of North America, covers different kinds of evolution: the big producers of the Middle East complain of being left out, while there is vigorous development in South America, not to mention the North Sea. It is evident that the intensity of drilling in the United States, which is traditional, though fluctuating according to the policies being pursued, is responding to a historic situation, a historic necessity: important reserves, but with these dispersed over a very large number of small or medium-sized fields and located relatively close to numerous insatiable centres of consumption. One may dream, trying to imagine what the result would be if the same effort, representing some 50,000 holes drilled annually, were divided up pro rata over potential resources. The map of world reserves and production would no doubt be transformed.

Coming back to reality, what are the results of yesterdays's and today's efforts, and what may be expected in the next few decades? As regards the past, reserves have generally grown faster than production, despite the irregularity of finds. This is reflected by the increase, on a world scale, of the ratio of reserves to production. Since the beginning of the 1970s the phenomenon seems to be going the other way

[5] Incidentally, here is an interesting case of positive feedback. By persuading oneself of the coming "end" of oil, one creates conditions that accelerate or can even create, the feared event.

or at any rate to be stabilizing. Annual reserves approximately compensate for production. For 1977 the results would seem better. According to *Oil and Gas Journal,* reserves exceed production (3 billion tons) by about 800 million tons (a result not confirmed by the competing *World Oil*).

The world value of the ratio of reserves to production hides major differences from country to country. This fundamental relationship between reserves and production calls for two comments.

First, at the end of 1977 the ratio of the world reserves of 88 billion tons to consumption represented stocks for nearly thirty years. However, this ratio would quickly halve (in eight years) to stocks for fifteen years if production and consumption continued to increase at a rate of 9 percent per year, as in the "Golden Age" (assuming no further discoveries or revisions). However, in practice, nobody thinks nowadays that such a rate will be maintained.

Second, more interesting is the fact that to shift the ratio of reserves to production from thirty to thirty-one years, that is to say, to gain a year's supply, it would be necessary to find annually *twice* as much oil as is currently discovered in the world. Another way to present the problem is to say that to increase world production by a billion tons while keeping the ratio of reserves to production constant (in reality, its progressive decline is inevitable) it would be necessary to add 30 billion tons to the reserves! Historically this is the phenomenon the world has known; indeed, it has done even better, because the ratio of reserves to production has increased over time, thanks to the "miraculous" discoveries in the Middle East, among other factors. The opposite has occurred in the United States, where additions to the reserves, even large ones, could not keep the R/P ratio constant, so that this has gone down to nine to ten years. Considering a ratio of ten as the minimum that must be maintained, if the United States wished to increase their production by 100 million tons per year, they would have to add a billion tons to the reserves, a task generally considered almost impossible, unless another Alaska is found. Another example involves the North Sea. Despite the valuable contribution of North Sea reserves to world reserves, in the short term they tend to *reduce* the world R/P ratio because they have been exploited so fast. (The R/P is about fifteen to twenty years, less than world average.)

From all this it will be understood that it is extremely difficult to evaluate the future rate of finds and/or additions to reserves, taking into account all the uncertainties about ultimate resources of oil and future exploration policy. At best, one may turn to the analysis of the consequences of certain more or less cautious hypotheses. This was done, independently but at about the same time, by Pierre Desprairies on the basis of the results of his Delphi poll on resources and by Carroll Wilson with his WAES group (Workshop on Alternative Energy Strategies) at MIT.

The former took the hypothesis that the history of world oil discoveries would evolve much as in the United States, a country with varied geology, intensively explored, currently entering a period of decline in the progression of finds, subsequent to an initial phase of exponential growth of these, followed by a period of linear growth. This idea of a phase of decline is incidentally violently opposed by some, including Hollis Hedberg, who take into account the potential exploration of new regions, including the off-shore Atlantic (which is taking place, though not without enormous difficulties and some disillusionment, as with the Baltimore Canyon).

The possible values for gross annual additions (revisions and discoveries) over the world have been calculated in the Delphi poll using three assumptions regarding ultimate resources: a low value of 175 billion tons, a middle value of 240 billion tons, and a high value of 375 billion tons, still to be produced.

In all cases, the gross average additions diminish fairly regularly with time. In fact, the experts who replied in the Delphi poll see the gross annual additions (new fields, plus reevaluation of old fields) rising from 4 billion tons per year in the near future — the 1977 and 1978 results seem to confirm this — and falling to 3 billion tons or less per year around 2000. In practice only 45 percent of these additions around 2000 will come from discoveries of new deposits, net, and 55 percent will come from the reevaluation of old deposits, partly due to having major recourse to enhanced recovery, a phenomenon that will take on increasing importance with time. Personally, and leaning towards the higher estimates, we would expect discoveries to be spread over a long period; that is, we would expect flatter curves (hat-shaped rather than bell-shaped) over a longer time.

Carroll Wilson's team took two possible values for the rate of additions from now until 2000 (for the Western world, excluding the socialist countries): 2.74 and 1.37 billion tons respectively (20 and 10 billion barrels per year, compared with 15 billion barrels average per year recently over several years in the Western world). The higher figure also assumes a relative decrease in net finds and a progressive increase in enhanced recovery. Between 2000 and 2025 gross finds would decline progressively to 0.55 and 0.41 billion tons per year respectively. It may be noted that the "high" hypothesis corresponds to about 110 billion tons of oil still to be discovered in the Western world, a hypothesis which comes fairly close to the median results of the Delphi poll. The "low" hypothesis, on the other hand — which the WAES members often associate with a slowdown of world economic growth becoming more or less permanent — seems relatively pessimistic: about 56 billion tons still to be discovered in the Western world in the next fifty years, on average hardly more than a billion tons per year! In fact, the two phenomena, low potential reserves and slow economic growth, are not necessarily linked.

CURVES OF FUTURE PRODUCTION

The stage is now set for calculating possible future world oil production. One may hasten to say that despite a certain heterogeneity in their hypotheses, the results of the two studies by Pierre Desprairies and Carroll Wilson are generally fairly comparable.

From the numerous curves arising from the Delphi poll, according to the hypotheses made for future rates of production for different regions (or control of the ratio of reserves to production), we have taken two (see figure 8-2) combining the average estimates of resources by the central group of experts (238-240 billion tons) and rates of production either as in 1975 or accelerated: North America "controlled" to ten years, Latin America to twenty-three or fifteen years, Western Europe to fifteen years, the socialist countries to twenty-six or twenty years, the Far East twenty-eight or twenty years, and the Middle East/Africa to fifty-four or thirty-five years.[6]

Maximum production, about 5 billion tons per year, will occur around 1990, dropping back to 3.5-4 billion tons in 2000 (about 0.5 to 1 billion tons higher than the current level) and to 2.5 billion tons in 2020. That is to say, without saying it, that the world oil crisis around 1985 will depend on the level of production adopted by the countries of the Middle East and North Africa (in practice, essentially the countries of the Arabian Peninsula).

What Pierre Desprairies puts so delicately is stated by WAES in blunter terms. While the Delphi poll is mainly interested in *possible* curves of production, leaving to others the job of comparing potential production and potential demand in detail, WAES essentially addresses the question of the adequacy of supply in relation to demand or the gap between the two which could generate a crisis. As well as technical considerations, hypotheses are taken into account relating to politics, economic growth, and world energy.

It seems to be admitted nowadays that economic growth and energy growth were undoubtedly overestimated by the WAES. In other words, because of the recession that the world is finding so hard to get out of and because of inflation, which, taking one year with another, continues to gallop on in most countries (throwing out all economic calculations), and because of energy and oil saving due to increased prices, the demand for oil could be lower than originally estimated for the next few decades.

The WAES has developed the thesis that the "crisis" could occur between 1980 — tomorrow — and 1988 — the day after tomorrow. This "message" from WAES, because of the renown of its participants, has had a very big impact, the more so because it has been taken up and amplified, for various and sometimes contradictory reasons, by most of the oil companies, the CIA, and numerous "sciensationalists" in the energy field, not forgetting the proponents of nuclear energy.

CRISIS OR "NON-CRISIS"?

It should not be concealed that in this field the only certainty one can enjoy is the certainty of being mistaken. If by chance one is right, it is often for different reasons than one had taken into account.

[6] Values of the annual reserve to production ratio (or "control").

With this reservation, an important one, our personal opinion is that these curves or forecasts are *pessimistic* (as we have said ever since they appeared). Current events, and the revision of certain opinions, seem to indicate that we are right.

Evidently the forecast crisis could come about at any time during the next few years if the Middle East countries, in particular the producing countries in the Arabian Peninsula, cut back their production considerably. Leaving this hypothesis aside, two factors should be taken into consideration:

- As stated, it is probable that energy demand in general, and the demand for oil in particular, has been overestimated (Carroll Wilson himself agrees today);
- At the beginning of the 1980s other oils should also arrive on the world market, from Mexico, Egypt, Malaysia, Brazil (admittedly just for the Brazilian market, but thus freeing imports), the pay-off from the programs of exploration and development started or given new impetus following the events of 1973-74. In fact this is the beginning of a wave underestimated by most forecasters: at $20-25 per barrel, oil becomes a raw material that many developing countries can hardly afford to import but are tempted to produce, even in small quantities. This oil of the future also becomes a choice target for most of the national oil companies created in recent decades: first, the Italian AGIP in 1924, followed shortly after by the Compagnie Française des Pétroles (with government participation and control). In 1950 there were seven; the first in OPEC (and preceding OPEC), the Iranian one, was created in 1951. At the end of 1977 there were fifty-one national oil companies, plus twelve partly owned or controlled by their governments.

Given the current world oil situation, these companies are more or less open today to "cooperate" with the international oil companies. This can be seen in Brazil, which has been opening up again to international exploration after a long period of categoric exclusion. Whether these efforts result in the finding of oil, as in Mexico (but only through national effort, by Pemex), in the midst of an oil boom, or as in Egypt (whose production should reach 40 to 50 million tons per year by the beginning of the 1980s), or the finding of gas as in Pakistan or Thailand, the impact of this NOPEC (Non-OPEC) oil on world oil production could and should make itself felt from the first half of the 1980s, extending through this period and possibly increasing.

The prospecting carried out by NOPEC is being encouraged by the international climate, which seems to be thawing a little after the great crisis of mistrust following the events of 1973-74 and the producing countries' seizure of control of their own oil destiny, when they disregarded contracts, which had ceased to have much meaning. For some time this new prospecting has interested the World Bank, which is considering getting involved in the financing of oil prospecting in the neediest Third World countries (rather like the parable of giving a fish or teaching to fish).

Without going as far as the optimistic forecasts made by the Irving Trust, for example, which foresees that the reduction of demand and the influx of this new oil will considerably reduce the role of OPEC in the world market and will bring Saudi Arabia to the sacrificial altar with a production hardly over 100 million tons in 1985, as against a maximum capacity of 550 million tons in 1978 (!), one can imagine continuous and reasonable growth of production up to the 1990s, with levels between 4 and 6 billion tons, followed by a certain plateau. The bell-shaped curve is not inevitable.

Allowing for the fact that gas production (which we shall look at briefly in the next chapter) accompanies oil production and is about half as high, there would be between 6 and 10 billion TOE or 9 to 15 billion TCE available to the world in the nineties, compared with global consumption of *all* commercial forms of energy amounting to 8 billion TCE in 1977. On the other hand, taking the average value of 5 billion tons of oil around 1990, a rate of growth of about 4 percent per year, some 50 to 60 billion tons would have been consumed between 1978 and the beginning of the 1990s. This represents between half and two thirds of currently known reserves. It also means that if we take the conservative estimate of 260 billion tons remaining to be produced, in 1990 there would still be some 200 billion tons, or 40 years of production at a rate of 5 billion tons per year, giving us time. But for what?

MARKET PENETRATION BY NON-CONVENTIONAL OIL

If the crisis or, say, an increasing deficit between requirements and availability should occur in the middle of the 1980s, it is evident that non-conventional oil will not change anything. The time is too short, and the necessary capital will not be available quickly enough to build 20 to 40 plants for mining and processing (at a total cost between 50 and 150 billion dollars!) to produce an additional 100 or 200 million tons of nonconventional oil per year.

The situation will change completely after a delay of ten or fifteen years or more. This opinion also emerges from the Delphi poll, in which it is supposed that nonconventional oil will be able to penetrate the market during the last decade of this century.

This global appreciation may be broken down by countries. In Venezuela, for example, an oil-exporting country with a medium level of population (about 12 million inhabitants), there is still plenty of conventional oil (although heavy, that is, relatively viscous) that can still be produced at an interesting price, before the Orinoco heavy crude belt is attacked. A recent study showed that the cost of extraction of very heavy oil could be of the order of 5 to 6 dollars per barrel, to which another 5 or 6 dollars must be added per barrel for upgrading, including the removal of sulphur and metals such as vanadium, and the reduction of the density so that the upgraded crude can be processed in the normal way in any refinery. The cost of production of $10 to $12 per barrel was not very interesting for Venezuela before the world price of oil reached $20 per barrel. The financial needs of this country remain significant, of course, but the rates of expenditure and investment cannot increase there indefinitely.

The situation is quite different in Canada. The major exploration programs have been rather disappointing as regards oil, though they have led to important discoveries of gas, unfortunately in difficult Arctic frontier regions. Fearing that its balance of oil and its balance of payments might deteriorate badly around the mid-1980s, in 1977 Canada decided to give a fresh legislative and fiscal look at its enormous dormant treasure of heavy oils (the Lloydminster and Cold Lake deposits) and, especially, tar sands (Athabasca, Wabasca, Peace River, and Buffalo Head). Following the start-up of the second installation for tar sands, Syncrude, in 1978 (6.25 million tons of oil per year) similar installations will follow progressively (a Shell project is just starting, for example). Between now and 1985-87 Canada expects to make up its oil deficit of about 25 million tons per year with the help of its tar sands and heavy oils, the cost of the oil produced probably being about $15-20 (in 1978 dollars) per barrel, approximately the cost of an imported barrel. This is perhaps where tar sands really take off.

In the United States, which has the biggest known reserves of oil shales, the situation is again different. Periodically, for over fifty years, as shown by Moody's quoted comments and his comparison of 1922 and 1977, the impending exploitation of oil shales has been expected. Periodically those expectations have been blunted for various reasons — political, economic, financial, and technical. Nevertheless, there is a feeling today that the goal is in sight, and American shales look like being the cheapest source of substitute liquid fuels for conventional oil (compared with synthetic crude derived from coal, for example). President Carter has proposed, in mid-1979, an ambitious program of synfuel development, including oil shales. Important progress has been made in *in situ* tecnhiques and modified *in situ* techniques (occidental type). The U.S. Navy is interested in shales to replace "reserves" taken for civilian use. The Department of the Interior has now authorized the start of work on two pilot projects on federal sites, the Occidental-Ashland and Gulf-Standard of Indiana (each project with about 2 to 2.5 million tons of oil per year), while other projects are being designed or are reemerging (such as the projects "Colony," "Tosco," "Sun," "Superior," "Paraho"). There are many favorable indications both for the oil shales of Colorado and for the oil shales of the eastern regions, which new processes might make worthwhile exploiting. These eastern shales have the advantages of being close to the areas of consumption and of less-limited water resources (for the processing of the shales and the restoration of the sites). The coming year could well be decisive.

Because there are numerous heavy oil deposits of heavy crudes, tar sands, and oil shales (even if less enormous than the American deposits) in the world, it is reasonable to forecast some progress in such exploitation as techniques become available and as needs make themselves felt, probably in the course of the 1990s. Brazil, for example, already has a pilot plant exploiting shales and is pursuing studies for commercialization on a larger scale. In Europe, France, as we have mentioned, is continuing to study the shales of the eastern Paris basin.

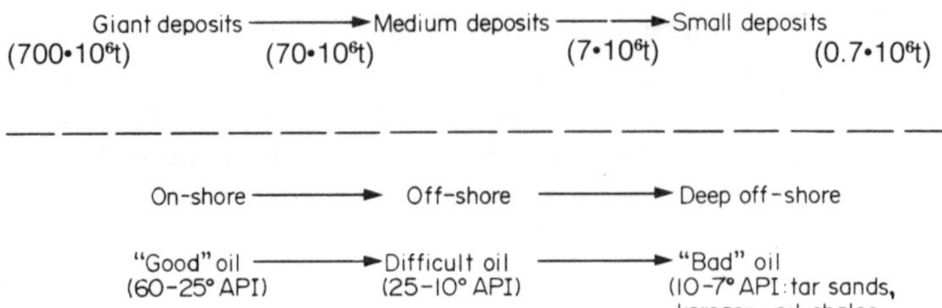

FIGURE 8-1. The Three Paths to ''Costly'' Oil

CONTINUITY OF OIL

Corresponding to the genetic and geological continuity of the various oil resources, one can speak of an ''industrial'' continuity, in which three ''lines of force'' may be distinguished. These lines of force go in the direction of increasing costs, though at different rates (see figure 8-1).

The first line of force goes from supergiant fields, or giant fields (the crock of gold, the target always eagerly sought by the international oil industry) to medium-sized fields (still good to have) and small fields (for example, the Paris basin: one takes what one finds). The supergiant oil fields are becoming rare; it is sometimes even asked if as many as, say, a dozen will be found in the future. The giants (half a loaf is better than no bread) are the goal of every optimistic explorer. These are currently the only ones being exploited in the North Sea. But remember that most fields in the United States (in terms of numbers, if not of production) are of medium or small size, a reason that the United States has long been, and still is, a country with expensive oil. It is also the reason for its important and permanent drilling activity. In numerous countries the absence of a national infrastructure and a national market of adequate size, the distance from major centres of consumption, and the fact that one could do better elsewhere have generally acted as brakes on research and exploitation of these medium or small deposits. The bolts have now been sprung, and this oil should play an increasing role in supplying a large number of countries. It is also ''approachable,'' with the necessary investment probably being below the level of $500 per barrel per day of capacity with the technical costs of production under the level of $5-6 per barrel. It is a question of being patient and waiting, the policy of the ''fine comb.'' It is also good business for the numerous national companies.

The second line of force is that going from oil on-shore to oil off-shore and deep off-shore. Technology has made and is continuing to make huge progress, in the North Sea and elsewhere; unfortunately, so have the costs! The investment varies considerably with the depth of the water (100 to 200 m or more, 350 m for the test in the Gulf of Mexico) and the state of the sea. In the North Sea, for example, it varies between $3500 and $7000 per barrel per day of additional capacity for the southern zone and $6000 to $12,000 for the northern zone, which gives technical costs of production of $3-6 per barrel or $5-9 per barrel respectively. In the deep sea, say below 300 m (though Shell is in the process of installing the Cognaq platform in the Gulf of Mexico, working at 350 m down), investments are said to be of the order of $15,000 per barrel per day of capacity and production costs of the order of $9-12 per barrel (1978 dollars). The size of the risks and the investments where the unit of account is a billion dollars (3-4 billion dollar projects are no longer rare) seem to reserve this domain to the happy few, first, the big international oil companies and a few of the biggest national companies, such as the Brazilian Petrobras with its Campos off-shore project.

The extremely difficult conditions found in certain regions of the globe, such as Alaska or the Canadian Arctic, in some off-shore zones such as the North Sea or in the ocean depth, rule out from the start all except giant or supergiant deposits, in short, exceptional deposits where the enormous investment necessary is justified by the size of the reserves *and* the productivity. It can be hoped that, as the infrastructure develops and is partially amortized, in time more modest deposits, possibly those around the major deposits, could also be put into production.

The third category of resources and the third line of force: from "good" oil to "difficult" oil, requiring enhanced recovery, for example, and "bad oil," the heavy oils, tar sands, and oil shales. At this end of the scale investments can vary from $5,000 to $10,000 per barrel per day of capacity for the heavy oils to $15,000

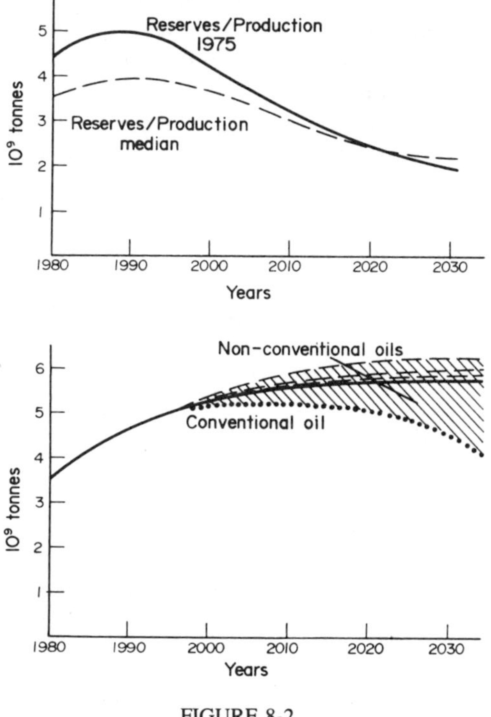

FIGURE 8-2.

to $20,000 or even $30,000 per barrel per day for tar sands and oil shales. The recent Shell tar sand project in Athabasca has been put at $28,000 per barrel per day of capacity in 1985. On the other hand, the Occidental project, using the modified *in situ* technique for Colorado shale, has been put at $8000 per barrel per day, a relatively low figure which justifies the increasing interest in this method. In this domain too, current projects cost several billion dollars and seem to be reserved, at least at the start, to the biggest companies (they themselves often getting together in consortia).

All these oil technologies form more and more of a continuum, in which the biggest companies, because of their financial resources, their know-how and technological and management superiority, and their taste for risk (particularly if it is generously rewarded) are progressively opening up new frontiers. In parallel, they are opening up new areas of resources, though at an increasing cost. In short, we see reflected here the progressive conversion to even costlier energy, but with relatively nondeclining abundance.

Insofar as no new form of energy shortly arrives, in the medium term and in large quantities, to undercut the price of conventional and nonconventional oils, we can imagine a composite curve for world oil production (the shape shown in figure 8-2 is more important than the figures). Such a curve looks very different from curves of the WAES or Delphi type. It presents continued growth of conventional oil up to the 1990s, followed by progressive penetration by nonconventional oils from then on, permitting the maintenance of a stable level of world production for many decades of "all oils," or even allowing slight growth over one or two decades more.

How to achieve these goals, which cannot be considered in isolation from an overall energy policy, will be discussed in our conclusions.

9

NATURAL GAS, OR, THE UNEXPECTED GUEST

Natural gas is rather like the "unexpected guest"[1] at the big energy banquet. It also, unfortunately, provides the saddest example of wastage — OPEC even calls it a "crime" — in the energy field: billions and billions of tons of oil equivalent have been put to the torch and will continue to be burnt at a rate thought to be above 100 million TOE per year.

The industrial and commercial history of natural gas began after World War II, in the United States. Because of its qualities: cleanness, low sulphur content before or after purification, no storage needed on the consumer's part, thus also eliminating handling costs, simple maintenance of appliances and equipment, and also because of an attractive price artificially kept low, natural gas has seen a spectacular development in the United States, where it has gained a position almost equivalent to that of oil. Today it is the object of a typically American political battle: the too-low commerical price for gas between producing and consuming states has greatly benefited the latter, situated on the east coast and elsewhere, and consumption has increased in an anarchic fashion: price control has engendered uncontrolled consumption. But these unprofitable prices for the producers have discouraged the search for new finds, have led to accelerated running down of reserves, and created occasional or local shortages. The producers are requesting to be freed from price restraint, as promised by President Carter during his presidential campaign, but the east coast states oppose this and are particularly influential in Congress. This is a veritable new war between the consuming North and the producing South (headed by Texas), whose interests in fact appear closer to those of OPEC than to those of the citizens of Boston.

In Europe, the natural gas boom is still more recent. After a slow start with discoveries at Lacq in France and in the Po valley in Italy in the 1950s, it took off with the discovery of enormous deposits in the Netherlands at the end of the 1950s and the beginning of the 1960s, hardly twenty years ago. The boom was given new impetus by discoveries in the North Sea, first in the southern sector (British fields)

[1] To borrow an image from Jean Fourastié, the famous French economist.

and then in the norther sector (British and Norwegian fields, including the big Frigg field). Following this boost coming from the North came the potential from the South, Algerian and Libyan gas, for which the way was paved by the completion of chains for liquefied natural gas, with sea transport via tankers (starting in 1964), and this was followed by the ''East wind'' of the enormous Soviet gas potential.

Thus in western Europe consumption of natural gas has risen from 12 billion m^3 in 1960 (a billion m^3 of natural gas being about equivalent to a million tons of oil) to about 200 billions today, a factor of over 16 in sixteen years, or an average rate of annual increase of the order of 19 percent over sixteen years. Of course, this exceptional growth rate was favored by the fact that natural gas was substituted for town gas in numerous existing networks. But it nevertheless remains true that these developments were even more spectacular than they were unexpected, as witness the energy plans of the 1950s, which were ardently hoping that the atom would bring them the support that in fact natural gas provided. The development of natural gas has been equally vigorous in the USSR, where consumption has risen from 45 billion m^3 in 1960 to more than 360 billion in 1978 (a formidable technical achievement, given the difficulties and the distances), and in some producing countries such as Algeria, or, more recently, Iran. It can be said in a sense that in this decade gas became a world power, with consumption and production in an increasing number of regions and countries. While in 1960 the United States alone accounted for 78 percent of the world consumption of 445 billion m^3, their share is now only 37 percent (which is nevertheless considerable) of world consumption, which by 1978 had grown to 1490 billion m^3.

FUTURE PROSPECTS

The ''youth'' of natural gas means that its problems differ from those of oil, apart from the physical and/or technical differences between the two products. While there is a well-established world market for oil, though not without its problems, international trade and, in particular, intercontinental trade in natural gas is in its infancy and has its own rules. To appreciate the prospects we must first of all take a look at the reserves and resources of natural gas.

Reserves of natural gas were estimated on 1 January 1979 to be 70,726 billion m^3,[2] about 71 million TOE, nearly 20 percent less than world oil reserves. The ratio of reserves to production is much higher, more than fifty years for gas as opposed to some thirty years for oil. These gas reserves differ from those of oil in two ways: they are growing rapidly and their geographical distribution is different from that of oil reserves, a further attractive feature of gas.

We should recall that there are two types of natural gas deposits; the so-called dry deposits, containing almost exclusively natural gas (with some condensable vapors called ''associated liquids'' or condensates), and the deposits that are in

[2] *Oil and Gas Journal* estimate.

fact oil deposits and are prospected for as such, containing significant quantities of "associated natural gas." It was this associated gas that used to be burnt and is still sometimes burnt with a torch in the absence of outlets. The oil-producing countries, including OPEC's members, are now making enormous efforts and putting enormous pressure on the operators to find outlets for this associated gas or at least to reinject it. We may quote the example of Iran, where, following an enormous program for the injection and reinjection of gas, the recovery rate for oil should increase from a measly 16 percent, due to the special characteristics of the chalky productive layer, to a good 30 percent, increasing recoverable reserves by nearly 30 percent and bringing them from 9 to 12 billion tons.

As for dry gas, unhappy have been those who found it in the past instead of oil. Even if today it sometimes fails to gladden those who find it, including those in the North Sea, because of the difficulties and costs of getting it into production and transporting it, the situation is nevertheless changing rapidly; and it is being sought more and more actively in numerous parts of the world. Indeed, more and more is being found, including at depths where no oil is to be found because of thermal conditions but where modern drills are able to penetrate in search of gas.

As regards the geographical distribution of the 70,726 billion m^3 world reserves (1/1/1979), the largest reserves are found in the USSR, with 25,753 billion m^3 (about 36 percent of world reserves). Then comes Iran with 14,150 billion m^3 (20 percent of the world total), whose future gas projects are today taking over from oil projects, with new oil seemingly reluctant to be found. The case of Iran also illustrates the extent to which natural gas formerly failed to attract interest: not so long ago reserves there were estimated at 1000 billion m^3. The jump to 14,000 billion m^3 resulted from discoveries (including dry gas), but also from reestimation of the real reserves of associated gas.

Then come the United States (8.2 percent of the world total), Algeria (4.2 percent), Saudi Arabia (3.9 percent), the Netherlands, whose stocks are running out (2.5 percent) and the North Sea (1.1 percent), whose southern deposits are beginning to run out but are being replaced by new northern deposits.

Although OPEC's gas reserves are significant, they cannot be compared with the dominant position of their oil reserves, *a fortiori* in view of the difficulties of making these commercial (to which we shall return).

What counts in any overview of long-term prospects is not reserves, but total potential resources. Regarding these, two initial comments may be made:

1. The problem for natural gas is relatively simpler than for oil, in that the recovery rate is already high, from 80 to 90 percent. Although some progress may be possible, there are fewer unknowns here and less incentive to improve the rate of recovery than there is for oil, where the promise of improved recovery should considerably change the final oil balance.
2. Apart from that, our knowledge of total potential gas resources is hardly better than for oil resources (particularly as interest was first aroused much later, and

the number of estimates made has been relatively more modest). And most of the comments we have made about oil are also applicable, strictly speaking, to natural gas resources.

With these reservations, estimates for final gas resources remaining to be produced have also been revised upward over the course of time, and these estimates vary quite widely.

A good value currently considered, and accepted by the World Energy Conference, is 270,000 to 280,000 billion cubic metres, about three times as much as assured reserves. If this value is compared with cumulative production from the very start (that is, in commercial production, for the amount simply burnt away is inestimable, in all senses of the word), it turns out that about 8 percent of the gas originally present has been consumed (as against 17 to 20 percent for oil), which well illustrates the youth and potential for the gas industry.

How are these world resources divided up? OPEC, because of Iran and in general because of the gas associated with its oil, wins by a short head with a third of the total, closely followed by the USSR, the world giant holding the individual title, with about 29 percent of the total (including Eastern Europe, which is comparatively unimportant). In other words, the USSR and OPEC together control about 64 percent of final gas resources. Then comes North America with 19 percent of the world total (but which has already produced 25 percent of its total resources in the ground since the start of consumption); as for Western Europe, its share of the world total would be about 4.4 percent.

The potential of natural gas, assured reserves and resources remaining to be discovered, is thus extremely important. As regards discoveries it may be noted that though it is becoming increasingly rare to find giant or supergiant deposits of oil, discoveries of very good gas deposits are becoming more frequent.

How can this potential be translated for the future? Most experts agree in believing that current production could at least double between now and the end of the century, and that sufficient natural gas would remain to maintain production and consumption at a level of 2500 to 3000 billion m^3 per year for fifty years, this level, incidentally, represents the current world level (in equivalent energy) of oil production and consumption. The countries or regions whose cumulative production is most important (such as North America) or where production is currently most intensive (like North America and Western Europe) might see their production reach a ceiling between now and the end of the century, while other countries or regions (like Latin America or the OPEC countries) might see their production increasing beyond the beginning of the next century.

There is currently no relationship between the price of gas and that of oil. This can clearly be seen in the United States, where the gap, which was more than a factor of three, is nevertheless getting smaller. The experts of the World Energy Conference have estimated that the totality of ultimate resources of gas could be produced if its price reached between now and the year 2000 a level of 20 dollars per barrel of oil equivalent. If the price were not higher than 14 dollars per barrel of oil equivalent, on the other hand, this would probably lead to nearly a quarter of the ultimate resources being left in the ground. (For example, is it not being said, in the face of too-low prices and better-known costs, higher than anticipated, that a North Sea Frigg would not be developed if discovered today?) What has happened in the United States in recent years may be symptomatic: annual discoveries have regularly decreased, from 283 billion m^3 in 1973 to 245 billion m^3 in 1976 (appreciably less than consumption), but rose to 392 billion m^3 in 1977 because of an increase in prices and in the hope that these would be completely deregulated. The 25 percent difference in production between 14 and 20 dollars per barrel oil equivalent naturally represents gas from frontier regions, off-shore, from the Arctic, and the like. Most final resources could certainly be produced at a *cost* appreciably lower than 14 dollars (1976 dollars) per barrel of oil equivalent.

Unlike oil, of which nearly two thirds is consumed in countries which do not produce it, as a little more than one ton of every two produced is part of international maritime traffic, natural gas is currently consumed for the most part by the countries producing it. Both now and in the past it has been a relatively ''domestic'' resource. Of the 1341 billion m^3 consumed in 1977, only 146 billion m^3 (11 percent) formed part of international trade, and most of this was limited to a purely regional basis (Netherlands to France or Germany for example). The share of maritime trade accounted for by liquefied natural gas (LNG), in tankers, is still very modest: about 15 billion m^3 in 1977 (1.0 percent of the total produced) mainly between Algeria, Libya, and Southern Europe, and from Brunei and Alaska to Japan (since 1969). The first few cubic metres of LNG arrived in the United States from Algeria in 1978 (a perpetually renegotiated contract between Algeria and El Paso).

Given the enthusiastic and general growth of natural gas consumption, including by the nations most lacking it — showing that they are not put off by the historical example of oil, or that their hunger for energy is insatiable — it might be thought that the situation is changing radically. In fact, it is not as simple as that. Interregional commerce in Europe and elsewhere is approaching its limits (sources: Netherlands and North Sea), unless the USSR is included, whose export potential may continue to grow.

The big question concerns maritime trade with LNG via tankers, and here the chains (pipeline from the deposit to the sea, liquefaction plant, loading, tanker, unloading, regasification, storage, distribution) are relatively rigid and particularly cost-intensive. They are so expensive that it has just been decided to build a submarine pipeline between Tunisia (sourced from Algeria) and southern Italy, and that serious study is being given to a second one between Algeria and Spain (Straits of Gibraltar), ultimately supplying France. Here, one might note the political importance of an energy link crossing between Europe and Africa, with these two pipelines being prime examples.

One of the most acute questions in the world's natural gas trade, which has still to be created, is the role of the United States. They are already the biggest importer of oil. Will they also become the biggest importer of natural gas, including LNG? Supporting this thesis is the *assumed* insufficiency of their resources and the real growth of their requirements. Opposing this thesis is the *potential* sufficiency of their resources and the growth of their energy dependence, which is now encountering the ideal, vigorously propounded, of Energy Independence (which Nixon hoped to reattain in 1985 with his famous Project Independence). Is there some contradiction as regards resources? No, for resources of conventional natural gas seem to be effectively limited, although the limit may be further away than it is for oil. But for the cost of importing LNG, other resources might be possibly tapped: gas in the geopressured zones in the Gulf of Mexico (*in situ* resources estimated between 72,000 and 123,500 billion m^3, of which 5-10 percent might perhaps be recoverable); gas in tight formations (15,000 billion m^3) or in Devonian shales (12-15,000 billion m^3), or methane, which could be recovered from coal beds before exploitation (7500 to 20,000 billion m^3), all these resources being apparently abundant but expensive to produce. This is not to mention the gasification of coal, with which the first pilot plants should or could be able to demonstrate their technological feasibility and, it is hoped, their economic competitiveness (an additional argument for raising the price of natural gas, so that synthetic gas might become competitive).

The existence of such additional nonconventional resources is problematic in Europe, as to date they have practically never been sought. On the other hand, Europe leads the field in studies on gasification, although its coal, much dearer from the outset, can hardly justify its use. OPEC also regards such projects with a jaundiced eye and thinks it better to invest in its abundance of gas rather than in such superfluous projects as the gasification of coal. Unfortunately, the 1973-74 oil embargo is one of the rare examples where countries seem to have a long memory.

Also, gas does not have only advantages, as some recent accidents show (in France, rue Raynouard and in Aubervilliers, in Spain near Tarragona with liquefied gas). It might be said that a similar accident would doubtless have sufficed to bring nuclear power to a definitive standstill, and in what confusion! These accidents have nevertheless refocused attention on the risks of living with such an explosive material, including in towns, which under ground are sometimes as full of holes as gruyère and can slip or subside locally, thus damaging piping systems. The concentration of LNG at ports of embarkation also constitutes a threat and has weighed heavily on American importing projects.

Despite these dark spots in the picture, it may be felt that because of its qualities in use and the abundance of reserves on a world scale, particularly since a significant share is *associated* with the production of oil, natural gas could have a future no less dazzling than its past.

10

LIGHT AND SHADOW ON COAL

The first chapter was entitled "energy demand: an unknown quantity." The last could be called "coal: an unknown quantity." Between the two, we have presented more questions than answers, more uncertainties than assurances.

Having been displaced by oil and pushed back to the last ditch (electricity, and even siderurgy, which has flirted for a while with the atom) by nuclear power, coal is today threatening to take its sweet revenge. In fact, the "threat" or promise is still but a pinhead, for as David S. Freeman, head of the Ford Foundation Energy Policy Project, has said: "The annoying thing about coal is that we do not know how to produce it or how to utilize it. . ." There is some truth in this quip, but not the whole truth.

The first trump card held by coal is its reserves and resources, which can be called immense! They are in any case very large compared with the estimated resources of *conventional* hydrocarbons or compared with our energy needs over the next few centuries. We must begin by asking what these reserves and resources are.

A DREAM OF ENERGY

At the request of the Conservation Commission of the World Energy Conference, in 1977 a group of experts proceeded to make a new estimate of world coal reserves and resources (using an international inquiry taking in all countries).

As regards resources, the figures are really impressive: more than 10,000 billion tons of coal equivalent[1] (more than 10 teratons) or, to be precise, 10,126 billion tons. (The precision is of course illusory, simply resulting from the addition of the various values stated.) This is nearly four thousand times the current production/consumption, sufficient to continue this, theoretically, without any additions for nearly four thousand years. These are evidently gross figures, and it is necessary to understand the subtleties they may conceal.

[1] That is, related to a standard coal with 7000 kilocalories per kilogram.

The first question one may ask is what these figures really correspond to, what kind of coal we are talking about. The experts note that these are indeed resources, the values given being completed by information on the depths of the beds (generally lower than 1200 m), their approximate thickness, and in some cases further supplementary characteristics. They are not referring to what is called ''occurrences,'' that is, zones where there is coal but about which more cannot be said. This figure of 10,000 billion tons is to be considered realistic. It is the aggregate for various types of coal, from anthracite to lignite or brown coal, from high to low calorific value (excluding peat).

These ''occurrences'' also merit some further comment. Unlike oil, which has been sought more or less actively for over a century, coal has never really been *sought*. In most cases it has simply been found, easily, because it had surface outcrops, and it was only necessary to follow the seam. This coal was found in such quantities and in the countries that needed it, in the process of industrialization,[2] that (unlike oil) it was not necessary to go and look for it elsewhere.

This is the reason that many developing countries do not *know* if they have coal; or if they do, they do not know how much. The door is open to exploration. Coal deposits, even small ones, in many cases would resolve some of their energy problems and their economic problems by utilizing a domestic resource that could employ generally abundant manpower. We often say, and there is some truth in the quip, that before building a nuclear power station in a developing country, or a solar plant, one should at least check whether there is no deposit of coal (or oil or gas) under the selected site!

The 1977 study was, of course, not the first estimate made of world coal resources, one such major effort dating back to the Toronto Congress in 1913. What is interesting is that such estimates in recent years have been revised *upward*. The estimate made for the World Energy Conference in 1974 was only 8,603 billion tons *(sic)* and in 1976, 9,045 billion tons. That is to say, in three years, with each country questioning itself about its coal after the 1973-74 oil crisis, estimates added about 1500 billion tons to the resources. This addition is higher than the total ultimate resources of conventional oil *and* gas given in the preceding chapters.

This enormous amount of coal does not merely exist on paper. It is also reality — not everywhere, indeed, and not enough in Europe, unfortunately. There are coal deposits, the dream of the energy expert, which almost surpass all imaginings: tens of metres thick, extending over several thousand square kilometres — for tens, maybe hundreds of thousands of km^2 in Siberia, with thinner seams, it is true — some only a few metres below the surface. The first time one is confronted with a ''wall'' of coal 20 to 30 metres high (the height of a ten-story building), in Wyoming, for example, one can hardly believe one's eyes, let along the possibility of an energy crisis!

[2] The question is sometimes asked: Is it countries in the process of going industrial that ''made'' coal, or is it the existence of coal that ''made'' the industrial countries?

Where is this manna, this coal, to be found? Alas, it is spread very unevenly. The number of countries with coal is certainly high, but the pyramid of treasure has a wide base and a very small summit.

At the head come three countries or supergiants where coal is concerned: the giant of giants, the USSR, with 4860 billion TCE, nearly half the world's resources. Then come the United States, with 2570 billion TCE, and China with 1438 billion TCE (these certainly being minimum figures. Incidentally the thickest seam of coal is found in China: 130 m, a skyscraper more than 40 storys high!). Together, these countries possess resources estimated at 8868 billion TCE, about 88 percent of the world total. Nature only lends to the rich or the great.

This category of the super-rich (coal resources over 1000 billion TCE) is followed by half a dozen very rich countries with resources between 100 and 1000 billion TCE. These are (in decreasing order of resources): Australia (262 billions), Federal Republic of Germany (247 billions), United Kingdom (163 billions), Poland (126 billions), Canada (115 billions), all industrialized countries, followed by Botswana (100 billions), last on the list. Although these resources are appreciated to a differing degree by the countries possessing them, and who think with some disdain that it is a long way from the "cup" in the ground to the "lips" of the user, it may be noted that each of these countries is sleeping (sometimes restlessly) on resources of coal whose calorific content is just about equal to the total world *reserves* of oil. That should wake them up a bit.

Five countries follow next, with resources between 10 and 100 billion TCE, South Africa and India (each with 57 billion TCE), Czechoslovakia (17.5), Yugoslavia (10.9, mainly lignite) and Brazil (10).

At the tail end of the coal-owning countries are nearly a score of countries with resources of between 1 and 10 billion TCE, that is to say, each having on average practically a "North Sea" in energy terms under its soil. These countries are situated almost all round the world, from Venezuela to Bangladesh, from Swaziland to Japan.

However, not all this coal in the ground can be considered a commercial prospect or as energy available. Under current economic conditions, geared to a guiding price of $12 to $13 per barrel of oil ex Persian Gulf, at the time of the survey, and with the technical means of exploitation possessed, about 6 percent of these resources are classified as reserves that are technically and economically recoverable. Six percent does not seem much, but 6 percent of 10,000 billion tons means about 640 billion tons of coal theoretically available, which is more than comforting: eighty years of consumption at the current world energy level, all sources included. The World Energy Conference experts assure us that these estimates, continually revised upward (473 billion TCE in 1974, 560 billion TCE in 1975, 640 in 1977) were probably conservative, so conservative that even doubling them to 1200 billion TCE would appear quite realistic.

With a few small differences, the world distribution of recoverable reserves more or less parallels that of resources. The USSR, United States, and China of course remain in the lead, with 110,178 and 100 billion TCE respectively (but note that the relative positions of the USSR and the United States are reversed).

THE PROBLEMS AND POTENTIAL OF COAL PRODUCTION

Except in Eastern countries, coal production is rarely proportionate to coal potential. This is why most countries are currently reexamining their coal programs, and mostly planning substantial increases.

Today ten countries share (with more than 2.2 billion tons) some 85 percent of world production, totalling 2766 million TCE (1978, according to the United Nations). These are (in 1978 figures) USSR (515 M TCE), the United States (590), China (496), Poland (203), the United Kingdom (124.7), Federal Republic of Germany (121.5), India (110), Australia and South Africa (84), and Canada (28). One should also add the German Democratic Republic (84.2) and Czechoslovakia (85), but their development potential is apparently much less than that of the ten countries cited. For these ten countries total production might reach or exceed 8.1 billion TCE by the year 2020, with some quite spectacular developments: United States, 2400 million tons alone; USSR and China, 1800 million TCE each; India, 500 millions; Australia, 400 millions; Poland, 320 millions; United Kingdom and Canada, 200 millions each; and so on. Total world production might exceed 8.7 billion TCE in 2020 (the limiting horizon of the World Energy Conference studies).

Some of these coal giants evidently will not consume all the coal they are capable of producing, according to these forecasts. That is to say that there could be a bigger world coal market than there is today (about 200 million tons in 1978, or 7.0 percent of total production, largely comprising coking coal for siderurgy), but which is hard to define or limit. For a market, there must be sellers (and apparently there are some!) and buyers. Now it is not clear at what point countries today importing energy — principally Japan and the northern and southern parts of Western Europe — will really be buyers of coal in very large quantities. Currently its price must be attractive: 30 dollars per ton delivered to a European port, equivalent to less than seven dollars per barrel of oil equivalent. The long-term trend of these prices is not well known, nor is it known when the good surface deposits now being exploited will begin to run out and when it will be necessary to return to underground recovery, which is much more expensive. The other problem is the attitude of current and future users: in the United States President Carter has imposed restrictions on the industrial use of fuel oil and natural gas, consequently obliging users to convert or reconvert to coal, which is happening, like it or not (mostly not, from what one can hear). In Japan there is an intelligent and aggressive program to bring coal back in stages. But in Europe, for the moment, there is nothing of the sort, and no clear outlook. On the one hand there are the nuclear programs, which it is hoped will get moving again one day, and on the other it looks as if the outlook for oil (and gas) supplies may some day appear brighter. The possible coal users are counting heads and asking themselves about the future, particularly as (except in the United Kingdom) they do not have the stimulus and promise of cheap and abundant domestic coal.

The possibility of a world coal market, which is interesting an increasing number of energy experts (including the oil companies, which could play a significant role thanks to their financial resources and their commercial ability on a world scale), raises another interesting question. Besides the merchant producers and small users, comparable in some respects to certain oil-producing countries, what might be the role of the three coal giants, the USSR, the United States, and China? What might be their share? As regards resources and technical capability, they could theoretically corner most of the market. Against this, there is their internal demand, large and growing vigorously, and the not always favorable location of their resources, far from ports or facilities for transportation abroad: the western United States on the one hand, Siberia on the other, several thousand km away from the east coast or the European networks. There is also another factor in the United States, the environmentalist opposition. While at a pinch it might modify its attitudes to prevent an economic or energy crisis leading to the collapse of the country, would it not oppose operations carried out solely for motives of commercial profit?

One may also ask if the concentration of a world energy market in the hands — the sometimes politically heavy hands — of the three political giants would be desirable — or desired by users already unhappy at having to depend on Saudi Arabia or Iran for their oil. Users would no doubt feel very small under the double umbrella of the atom and coal held by their all too powerful protectors and suppliers or having to depend on a handful of suppliers, whether regrouped or not in the bosom of an "OCEC," or Organization of Coal Exporting Countries.

PRODUCTION: FROM THE SHOVEL TO UNDERGROUND ALCHEMY

The most spectacular development in world coal production in recent years, and the guarantee of fabulous growth in production discounted on the future, is the "return to the surface" of coal mining, that is to say, open-cast mining or quarrying. This development can be summed up in one word: gigantism.

There has been a real technological switch in coal production: the miner has swapped his pick and the modern machines that followed it for elephantine public works machines: 100 m^3 shovels, giant mechanical draglines as high as a ten-story block, bucket-wheel excavators "eating" 200,000 m^3 of coal or overburden a day, conveyor belts, armadas of lorries with 80 m^3 skips, and the like. The production capacity of open-cast mines using these "mechanics" of the secondary era has risen from 10 to 20 million tons per year per mine towards figures of 50 to 100 million tons per year, in Siberia for example (the Kansk-Achinsk basin). The record for this class seems to be held presently by West Germany, with the Garsdorf lignite mine near Cologne: 50 million tons of coal per year with a maximum depth of 300 m (bigger than the Eiffel tower!). The super record in the same region is already being prepared for the Hambach mine, whose production by the end of the century should reach 100 million tons per year, with a maximum depth of 600 metres!

Are these spectacular achievements the ecological catastrophes they are sometimes made out to be? It is evident that open-cast mining enjoys, often rightly, a bad reputation: polluted water courses and groundwater, earthslips, countryside despoiled forever. These bitter fruits of wild capitalism, all too frequent in the past in the United States, are not however inevitable, as can be seen from the lignite mines in the Rhine valley. To put the problem into perspective, let us first remember that the production of some 100 million tons of lignite per year entails temporary "devastation" of 240 hectares of agricultural land *per year;* urbanization, industrial expansion and the construction of roads and motorways, in the same region, "consume" definitively about 120 hectares of this same agricultural land *per week.* Be this as it may, the 240 hectares annually disturbed by mining are the subject of meticulous programs of restoration that give back to the countrymen or displaced inhabitants (this displacement greatly contributing to the costs of the operation, as one may imagine) land at least equivalent to that of which they were temporarily deprived. Furthermore, this land is integrated into a modified, even improved landscape with more marked features and with a few lakes, popular recreation areas. This agreeable situation, the result of long experience and a deliberate program of rehabilitation of disturbed land, is not of course general, perhaps because the exploiters of the land did not care — as most frequently in the Appalachians in the United States and elsewhere where the countryside is sadly disfigured — or because the ecological and climatic conditions are not as favorable as in the Rhine valley. Such is the case in the American West and elsewhere, although there are degrees of difference. In Wyoming, for example, the soil is poor, covered with poor grass and scrub; on average it needs 12 hectares to feed a cow. It may be noted that, given a seam of coal 15 metres thick, this cow may be weighed against 2 million tons of coal!

There is not much water, but the restored land compares advantageously — we have seen this personally — with its former state. In New Mexico, on the other hand, the soil is still poorer and water still rarer: every drop (from the Colorado River, for example) is planned for, and these new forms of utilization — restoration of the soil and/or process water for the treatment and conversion of coal — have to contest the charge that they are extravagant of water. The situation is still more complicated by the fact that these marvellous deposits of coal (and uranium and oil) are situated under land given to the Indians (thinking it was empty!). These Indians, widely dispersed, today use almost only the most rudimentary forms of solar energy, while having beneath their feet one of the richest energy treasures on earth, but no water.

These spectacular developments in open-cast mining, with a yield of about 40 tons per man per day should not lead us to forget traditional underground production, making laborious but constant progress, still supplying most of the world's production. Safety is making great progress, though at the cost of yield. Mechanization is almost complete in most countries, including long-face mines and moving hydraulic props in Western Europe, and this mechanization also extends to underground transport of coal on conveyor belts. These techniques are more or less reaching a plateau, while growing depths, often 1000 to 1200 m, have made costs grow exponentially, still burdened by manpower, the work being a little easier perhaps but still involving a relatively large number of men underground.

If the utilization of coal developed in accordance with the plans just mentioned — doubling between now and the year 2000, tripling or quadrupling between now and 2020 — the question of future types of underground mining would again emerge once the good, rich surface deposits had been mined and exhausted. But how? Not much progress can be expected with traditional techniques. New techniques, on the other hand, are still only in the experimental stage. Where is the manpower, for the work underground *remains* difficult?

A brief review will evidently not permit us to go deeply into these questions. Our object is just to draw attention to them, so as to be in a position to follow possible progress. Let us briefly mention hydraulic cutting, where the coal is "exploded" by water cannon (jets of water at a pressure of 60 to 120 kg/mc^2) with the cut coal transported hydraulically; about 2 percent of Soviet coal is produced in this way, and a mine in Germany has just started up using this method, which is well suited to contorted or moderately dipping seams. Another method being tested is robotization, where automatic machines controlled from underground would be replaced by robots with more degrees of freedom, possibly remotely controlled from the surface and with TV surveillance. These robots, resulting from the atomic and space programs, are still extremely expensive and have not developed as had been hoped at the time of the Apollo project or of advanced nuclear programs (the atomic rocket, for example).

There remain the possibilities of nonmechanical attacks on coal underground, attacks by combustion *in situ* or underground gasification, attacks by chemical solvents, or even microbiological attack. These last two techniques, the chemical and the microbiological, may be interesting in the long term; they are being studied in this spirit, that is, not very systematically.

The spearhead of this underground alchemy currently is *in situ* gasification. Strictly speaking, it is not really a new technique, because it was first used in the USSR in the 1930s, and two electric power stations have been supplied there since the 1950s with gas thus produced. The reticence of the Russians concerning these techniques suggest that they have not lived up to their promise: production would be irregular both in quantity and in quality (gas of low calorific content at 1000 kcal per m^3 instead of 9000 or 10,000 as with natural gas, and nontransportable, at that) and there would be fairly big leaks in the ground, which could amount to 20 percent of production (evidently a safety problem). The world renewal of interest in coal would have led the Russians to take it up again and rethink their experience. Major programs have started elsewhere, in the United States (an experimental program in Wyoming), in Alberta in Canada, and in Texas (under Soviet license!) for the enormous lignite deposits there, and in Europe.

In Europe, the Belgian INIEX (Institut National des Industries Extractives) has launched the interesting idea of pressurized gasification at great depth, which could increase the yield and especially the quality of the gas produced, also reducing possible leaks and risks associated with subsidence as well as the risk of interaction with underground water resources. A joint Belgo-German experiment is planned in the Borinage district of Belgium, while the French, who have belatedly become interested in this technique and seem to have sulkily rejected the offers of their Belgian cousins, are thinking of going ahead with an independent experiment.

There are plenty of technical unknowns with such a method, and among the economic unknowns are the drilling costs, but there is a lot at stake. First, such exploitation involves no personnel underground,. Second, if underground pressurized gasification at great depth proved feasible, this would make accessible a completely new range of coal requirements, *resources not currently taken into account* because they are below the normal depth limit of 1000 to 1200 m. In France, Belgium, and Germany, for example, several tens of billions of tons of coal could become accessible in this way (with recovery of the order of 60 to 65 percent, to judge from the results of American experiments).

Because of its low calorific value, the gas obtained with current processes is difficult to transport, which inflicts a big economic penalty. In Europe, with its great population density, its transformation into electricity at the pithead would not seem to present many problems. In the United States, on the other hand, and worse still in the USSR, the enormous distances are evidently a handicap. This is the reason why the USSR has a major complementary reserach program for the transport of electricity over very long distances (studies of transport of dc current up to 2 million volts).

The problem of transport is not a handicap only for gas produced by underground gasification. Coal itself has transport problems, and the poorer the quality, the greater the problem (as with coal from the American West or Soviet Siberia). While waiting for "carboducts," which sound promising (provided there is enough water) but are difficult to start up, the transport of coal over long or very long distances is being taken over more and more by special trains ("unit-trains"), often of 100 trucks with 100 tons (or 10,000 tons of coal per train) pulled by two or three locomotives. The unsatisfactory situation of having to return empty has led imaginative Texans to suggest transporting household refuse (from Houston, which doesn't know what to do with it) to Wyoming (where the coal comes from) and to use this refuse to backfill the quarries opened up by coal extraction. So that nothing is lost, the refuse would be allowed to "ferment" to produce biogas one day, as partial reimbursement for the coal debt.

THE POSSIBILITIES OF USING ALL THIS COAL

Shall we return one day to transatlantic steamers, or even to steam trains? Some British experts have suggested this, using boilers with fluidized beds.

Meanwhile there are essentially two classical applications for coal: siderurgy and the production of electricity. In the long term there are two very important potential applications: gasification to produce a synthetic gas of high calorific value as a substitute for natural gas and liquefaction or production of synthetic liquid fuels. At the centre of this square is an unknown: direct industrial use of coal in "improved" boilers, for example, using fluidized beds, both efficient and clean (combustion accompanied by removal of sulphur). In all cases coal has a handicap from the start: the cost and the difficulty of handling it, because it is solid. Its low price and its possible abundance might (must) tip the balance in its favor (perhaps aided by a "helping hand" via legislation!).

As for siderurgy, there is little to say. There are currently difficulties, including in some countries whose furnaces are greater than their needs. On a world scale and in the long term siderurgy should continue to grow, always providing an outlet for coal, which here is irreplaceable.

As regards electricity, there are numerous uncertainties. The electrical managers, who abandoned coal for more modern nuclear energy, the harbinger of the future, over the last decade would basically like to remain loyal to nuclear power, although it has lost a number of its economic advantages over coal. The first problem they have to face is that the growth of electricity is itself questionable, at least in its nonspecific applications and also in the heating market that it had so vigorously attacked. The second problem is to define, within a scheme in which growth may be slowing down, what will be the respective roles of nuclear power, coal, and other new sources of energy in the production of electricity.

A factor in favor of coal is that units of medium power, of the order of 600 MW(e), might come back into favor; another would be, as mentioned above, success in underground gasification, which would almost ineluctably lead to conversion into electricity at the pithead.

There remain the mystery and the promises of the conversion of coal into synthetic fuels, liquid or gaseous, in "coal refineries" or coal complexes. The principles are known, and pilot plants are in operation. Provided that the need makes itself felt, it is now only a question of financial and industrial mobilization, and of price.

It should be remembered that the manufacture of synthetic fuel was already carried out on a large scale during World War II, in Germany, and South Africa today is having recourse to it on a by no means negligible scale (the Sasol I plant, in operation since 1955, and the Sasol II plant currently under construction, designed to produce 2,100,000 tons of oil products annually, including 1,500,000 tons of gasoline, while consuming 40,000 tons of coal a day).

In brief, the object of these complex operations is to add hydrogen to coal — which already contains some, but not enough — to make it into a hydrocarbon.

While coal contains on average between 0.8 and 1 atom of hydrogen for each atom of carbon, hydrocarbons contain between 1.7 and 1.8 atoms. The source of the hydrogen used is water, often in the form of steam. It is a cheap source as regards the raw material, but costly in application. For the time being hydrogen remains expensive. The synthetic fuels that it will permit to be obtained from coal will also be dear, all the more so should the initial raw materials of coal also be dearer, which will unfortunately be the case for European coal.

The four basic processes used for the transformation of coal into hydrocarbon are: carbonization (or rather, hydrocarbonization), hydrogenation, extraction partial or complete), and the familiar Fischer-Tropsch method of synthesis (giving methanol, gasoline, and the like). The thermal efficiency varies according to process between 55 and 70 percent, and the material yield is between 180 and 600 litres of liquid fuel, plus between 60 and 300 m^3 of synthetic gas per ton of coal. To give an example, a coal refinery, probably using a combination of two or three basic processes, and using 25,000 tons of coal a day, could produce about 2,500,000 tons of liquid fuel and 2 billion m^3 of synthetic gas per year. The investment (not including the mine) would be over 2 to 3 billion dollars, and according to some recent estimates (perhaps pessimistic) the oil produced would come to 25-30 dollars per barrel, may be as high as 30-35 dollars per barrel. In mid-1979, such projects nevertheless received a strong push from the Synfuel program proposed by President Carter.

THE DANGERS OF COAL

As economic competitors for the production of electricity, are nuclear power and coal also competitors as regards the "apocalyptic" dangers they might bring upon our society?

Everyone knows that coal is "dirty" to use: with its NO_2, its SO_2, (found in acid rain) its dust and its soot, containing carcinogenic benzopyrene, traces of 14 or 15 heavy and dangerous metals, and even its radioactivity (said to be higher than the amount released into the atmosphere by a nuclear reactor of equivalent power), not to mention the 500,000 tons of ash, or thereabouts, produced by coal per year per power plant. All that would be quite "benign" alongside the threat presented by CO_2, which is such a terrible scourge that one may wonder if it is not the proponents of nuclear power who have conjured it up, or at least exaggerated it.

In fact, Callender pointed out over forty years ago that the concentration of CO_2 in the atmosphere seemed to be increasing as a result of the growing combustion of coal (and hydrocarbons). This could result in a long-term increase in temperature, which would be beneficial, extending, among others, the zone of cultivated land northwards. Today, after twenty years of measurement (from Hawaii to the South Pole), it can be confirmed that the concentration of CO_2 is indeed increasing, at about the same rate as is the consumption of fossil fuels, though variations in the biosphere and such phenomena as tropical deforestation also exercise an influence difficult to quantify.

Imperfect models (too simple, and generally without feedback), have shown that if the concentration of CO_2, currently 330 ppm, were doubled, an increase of temperature of 1.9° to 2.9° C *could* result in the troposphere, and even two or three times as much in high latitudes. The complex atmospheric circulation would be modified, in turn modifying local climates, including rainfall. There could be considerable consequence for local economies, which are built more than one may think on a certain climatic structure and constancy.

What should one really believe? An objective study of work in progress *points* to increasing evidence, but this may not be decisive for several decades. Should we act, or react? Without hesitation, some priority must be given to these researches (still very imperfect), together with the corresponding means. But what else to prepare for counter-action as regards climate? It seems a long way off. Intervention in the CO_2 cycle looks more probable; coal, like nuclear power, would then need a "fuel cycle" (note that the CO_2 recovered would be a possible raw material, though a "difficult" one, used for the synthesis of artificial hydrocarbons).

There are four "dustbins" for CO_2: the stratosphere, which has low capacity; the atmosphere, in which about 54 percent of the CO_2 already produced is concentrated and whose limited capacity is the cause of current concern; the upper layers of the oceans, whose exchanges with the atmosphere would also be too slow; and finally, the ocean deeps, with enormous capacity, an ideal "dustbin" on condition that the CO_2 got there. Hence the idea of "CO_2-aducts" ending at Gibraltar or the Weddell Sea, rare zones representing a vast ocean mixer-blender.

Another solution, more radical, would be not to burn any more coal. While awaiting the results of more urgent studies, such a solution would be just as premature as the opposite solution would be imprudent, namely, to decide today on an enormous increase in the world production and consumption of coal for a long period, among other countries, the United States, the USSR, Western Europe, and Japan, whom certain international ecologists are already accusing of having "poisoned" the atmosphere.

CONCLUSIONS ON THE PERENNIALITY OF COAL

To the first trump of coal resources (their abundance) we would add a second: their potential for conversion. Today it is being discovered that they can be transformed at will into almost anything, a property shared more and more, in fact, by all the members of the big family of fossil fuels: from coal one can make gas, from gas one can make methanol, from methanol one can make gasoline (a Mobil pilot plant is under construction), and so on. This potential for transformation has been summed up in the phrase: "The Arabs have succeeded in making the chemists think."

These transformations are or will be expensive. They must be compared with other energy solutions. On this comparison, also containing a degree of sentiment, will depend the future of coal: a servant remaining in the background or an old dictator, recalled and rejuvenated, as it seems we would like it to be today.

CONCLUSION: TOWARD AN ENERGY POLICY

It is clear that our technical civilization needs to redefine its energy policies, including its oil policy and its nuclear policy, as well as their respective roles.

LOYALTY TO FOSSIL FUELS

Let us hope that fossil fuels bear us no ill will, for they remain indispensable. Time has made them, and it is time that they can give to us.

Our civilization is built on coal, oil, and gas to such an extent that they have almost become second nature. We know how to produce them and utilize them, and they still account for more than 90 percent of our energy balance. Ignorant as to the resilience of our system in the face of too rapid changes in our energy structure, we are obliged to act with prudence. To increase our consumption of coal, oil, and gas is to inflate a current system that we know relatively well. This does not mean that the fossil frog of today may blow itself up to the size of the nuclear, thermonuclear, or solar bull of tomorrow; but the passage from fossil fuels to their successors must be progressive, considered, and controlled.

If one takes the figures from the World Energy Conference and adds the potential production capacity of fossil fuels up to 2020, with corrective additions for nonconventional oil and gas, one finds that mankind at this date would dispose of some 16 to 20 billion TCE of coal, oil, and gas, plus "a few" billion supplementary TCE from water power, nuclear power (thermonuclear power being at best in its commercial infancy at this time), geothermy, and solar power. This calls for two comments:

1. It is not an energy famine that threatens us. Of course, this situation could arise if we failed to take the necessary decisions in time, but it is by no means *inevitable*.
2. The role of fossil fuels will remain basic. Although they are doubtless relatively in decline, they will have assumed the role of permitting a smooth transition and providing a reserve for "noble" uses.

Nothing better illustrates this idea of a reserve than coal, ready to take a new direction. Today it is rather like the spare wheel on our coach of energy, and possibly it will be its motor tomorrow, over a long period of transition. The lens of our camera for exploring the future of energy in the short and medium term stays focused on hydrocarbons, especially oil, the role of which remains basic.

The producing countries are becoming wiser and more realistic. They have realized that the quadrupling of prices, violent as it was, was less brutal and less harmful than the shock of a collapse of confidence, following the noise of contracts being torn up. Their power is challenged by the possible return of a certain abundance of oil (an old, old story) following an economic crisis that neither they nor anybody else foresaw but toward which they contributed and which they helped establish, giving it the appearance of an energy crisis. Apart from tearing up contracts, the producers played the double game of OPEC and OAPEC,[1] only realizing too late that they were swapping an embargo illustrating their new power for a real desire for energy independence that had previously existed only in official speeches. As regards this game, one may well ask in whose court is the ball? Is it in the producers' court? In the court of the consumers and the companies (whose interests are linked, even if sometimes opposed)? Or is it in the net representing inability to produce oil as required, equally needed by consumers and producers?

The dialogue *must* be resumed, confidence must be restored. The consuming countries hold the key of technology, the consequence of grey matter exercised over centuries, and a wealth of riches: and they know how to use both. Technology cannot easily be transferred: it must be shared, and part of the road must be covered in tandem: "I give you a coin, you give me a coin, then we each have a coin. I give you an idea, you give me an idea, then we each have two ideas." Those who have ideas (which are renewable) must understand, appreciate, and *do justice* to those who have oil (which is nonrenewable). A partial payment or "reimbursement" in terms of solar technology could be a decisive element in renewing the dialogue and in the reestablishment of confidence. Europe would be particularly well placed to initiate such a strategy, both because of its relations with certain producing countries and because of the potential and the quality of its solar experiments.

Sometimes the question is asked: is it necessary to save oil? Certainly it is necessary to stop wasting it, but is it necessary to go further and to dispense with certain uses, prematurely perhaps? In the light of our chapters on fossil fuels we do not think this question is fundamental, because it is possible to prolong over an extended period the use of oil by using nonconventional oils, that is with the equation: oil or future liquid fuels equals new resources (shales, coals, and the like) plus technology — in constant progress — plus time, plus money (alas!). On the other hand we do not believe in the danger sometimes expressed that a new source of energy could develop so quickly, as was once thought concerning nuclear power, that the producers would remain with their liquid treasure unused, not in

[1] Organization of Arab Petroleum Exporting Countries.

their hands but under their feet. This might be true of coal (where there are immense deposits which would not be exhausted for many generations) or for "marginal" deposits of oil, or occurrences, but this will certainly not apply to the majority of oil resources, the advantages of which will continue to be appreciated for a long time.

Another question concerning oil is the future of the big oil companies, incorrectly termed international (for they are national , with only Shell being an interesting exception). Must they be suppressed; must their (dominant) role become a thing of the past? Should they be divested, vertically or horizontally? Vertically, that is, destroying the integration that was and is their strength and splitting them up into an independent network of producing companies, transport companies, refinery and distribution companies? Horizontally, that is, cutting the tentacles they have grown over coal, uranium, geothermy, solar power, all energy sectors (making them Great Energy Companies overall)? Or in such sectors as fats, ores, photocopying, or packaging, sectors which are generally highly profitable, where they can make use of their enormous cash-flows and their management know-how, taking an option on the future? You never can tell!

This battle of divestiture and diversification has its ups and downs, especially in the United States, where it is but one more manifestation of the indecision of the powers that be. It is really just a sham to amuse or mislead public opinion and delay the attack on the real problems. The real battle is and remains the battle of energy. It can and it must be won. We are preparing for it in a strange way — we live in a masochistic civilization — breaking on the altar of demagogy some of the best weapons we have to fight with. This is a personal deep-felt conviction, and not a plea to make these same companies the satraps of the energy world. They must be adapted, while allowing them to continue operating; these active forces must be channelled but they must be allowed to play their role (in oil and other energy sectors), since their potential and their know-how remain considerable. It would, for instance, be desirable to regulate the exploration sector to achieve a better knowledge of resources, a basic goal.

These companies have also contributed and continue to contribute usefully to a certain "reunification" of fossil fuels: geophysical and seismic exploration of all resources; open-cast mines for coal, tar sands, and oil shales; combustion *in situ* (fracturing[2] techniques, too) for improved recovery of heavy oils, shales, or coal; transport by pipeline; refining of oil shales; and liquefied coal. We currently have quite an armory of techniques, methods, know-how, and even manpower. All this gives the impression that the real age of carbon[3] is beginning today, the reply and the saving side-step in the shadow of new sources of energy, as if the traditional forms of energy also shared "the great fear of the nuclear era."

[2] Also useful in dry rock geothermy.

[3] It is proposed to substitute for the classification of fuels as solid, liquid, or gaseous the classification by means of exploitation: mining, drilling (an example of underground gasification, etc.). This is a more technological classification.

Because they permit better utilization of fossil fuels we should not forget energy-saving measures, a vast "deposit" only just beginning to be exploited, where the rules of the game are just beginning to be discovered. And the size of this deposit is just beginning to be realized by a world that has become wiser as to the growth of energy.

By thus satisfying the major part of our energy needs in future decades, fossil fuels, thanks to their relative abundance, will help us better to organize the transition toward new forms of energy. Let us hope it will occur without excessive haste and without over-hasty decisions, regrettably almost irreversible. Above all, fossil fuels give us an indispensable respite during which we can better weigh up nuclear and solar power.

THE WEIGHT OF NUCLEAR POWER

What nuclear power must suffer from is its weight, its extent, the decisions taken twenty years ago in other circumstances, for other conditions, and also before the birth of anxiety and public opposition that could not have been foreseen by the proponents of nuclear power and to which they have not been able or have not deigned to reply. Thus it appears as the final stage of a process of concentration and increasing complexity, which began two or three centuries ago. It can be said in a sense that the fast breeder reactor was not born ten or twenty years ago, but two or three centuries ago, when the machine of progress was set in motion, moving in a direction that became fixed and has not been questioned until very recent years.

The paradox of today's energy structures is their inertia; this compels decisions to be taken twenty, thirty, even fifty years in advance, in a world where our inability to foresee the future and change is notorious. Every society involves itself with the future, without in general knowing much about when or how. With nuclear power a little more is known, and this leads in the direction of anxiety.

KEEPING OPTIONS OPEN

It seems increasingly necessary to reintroduce a spark of imagination into our policies, including our energy policy. Excessively monolithic solutions militate against this. Rather than arrange a nuclear marriage for future generations in the Indian manner, where the parents arrange the marriage at birth or shortly after, our most urgent task is to open up as many options as possible and to eliminate as few as possible. We must make use of the time we have gained to explore the potential not of "soft technologies" but of "flexible technologies." In this respect, once again nuclear and solar power appear to be diametrically opposed. This apparent dichotomy is doubtless too simplistic, and the problem merits study in far greater depth. Obviously, the idea of keeping the options open does not appeal to decision-makers, who have little love for the probable and even less for the possible.

What conclusions could be drawn, given better comprehension or better comparisons between solar and nuclear power, following intensive considerations of the questions of energy during which such questions would arise? Doubtless, the conclusion would not be to choose a single, exclusive form, a mono-energetic structure: this will not be necessary for several centuries, if ever. Rather, we envisage a polyenergetic structure, where many sources and resources will have a part to play, leaving practically no stone, or mineral, unturned.

In practice, there will be two dangers to avoid. On the one hand there is that of adopting a solution (such as symbolised by nuclear power) geared to terawatts, whose rigidity could represent the most vulnerable factor to society in future. On the other hand there is the risk of foundering in a wealth of interrogation, leading to a collective, ''democratic'' paralysis with a sterilising effect, the prelude to the collapse of society as we know it.

Our mental effort should throw some light on what it is that society actually wants, on what nature can provide, and what the ecosystem can tolerate. Several ideas in this book may extend the time we have, but a comprehensive science of energetics has still to come into being.

Michel Grenon
Vienna, Austria;
En Meysconié, Tarn, France

POSTFACE

A year after having written this book, I have to say that we are not using as well as we should that resource I called our most valuable; time.

The dialogue so necessary between the consuming countries and the producing countries is still in its infancy. Events in Iran have dramatised the situation still more, if that was necessary.

The producing countries are becoming increasingly more orientated toward control of production, while the consuming countries, despite the pledges at the Tokyo summit meeting, are being forced willy-nilly to control their consumption, and worse still, to put controls on their purchases of oil. Direct oil sales by the producing countries have increased dramatically, and world prices are peaking in a disorderly manner. As for the consuming countries, they have got entangled in their own economic policies. In Europe, for example, in order to avoid hitting the dominant field of industry, with all the consequences that would ensue, the motor car continues to enjoy priority. Seated at the wheel, the consumer forgets all consciousness of the world. To stay there he is prepared to pay anything!

Now, at the beginning of 1980, the building of nuclear power plants has come to a stop in the United States, as a consequence of the Three Mile Island accident; it is progressing extremely slowly in Germany, Sweden, Switzerland, Japan, and the United Kingdom. It is practically only France that is pressing on undeterred.

Can the New International Economic (and Energy) Order be born of such apparent confusion?

The most important factor is that the new oil prices, 30 dollars per barrel today, perhaps 40 or 50 dollars tomorrow, will make all or nearly all solutions possible. New producers (NOPEC or Non-OPEC) *must* appear on the market; frontier areas, from the Arctic to deep-off-shore, are becoming economic. Tar sands, bituminous shales, and even gas or synthetic liquids derived from coal are picking up points. It is these factors, in conjunction with certain political factors, which justify the Synfuels program envisaged by the United States and, let us hope, by other countries in the near future.

In this sense, and despite the feeling that we are still in an alarming situation, the range of energy options is even more open than before.

Vienna, December 1979